GENES AND GENDER
FOURTH IN A SERIES
ON HEREDITARIANISM AND WOMEN

Series Editors
Betty Rosoff and Ethel Tobach

Gordian Science Series

Genes and Gender: I
On Hereditarianism and Women

Genes and Gender: II
Pitfalls in Research on Sex and Gender

Genes and Gender: III
Genetic Determinism and Children

Genes and Gender: IV
The Second X and Women's Health

GENES AND GENDER: IV
THE SECOND X AND WOMEN'S HEALTH

Edited by
Myra Fooden
Associate Editors
Susan Gordon and Betty Hughley

GORDIAN PRESS
NEW YORK
1983

GORDIAN PRESS, INC.
85 Tompkins Street
Staten Island, NY 10304

First Edition

Copyright © 1983 by
Ethel Tobach and Betty Rosoff

All rights reserved. No part of this publication may be reproduced, stored in a retrieval system or transmitted, in any form or by any means, electronic, mechanical, photocopying, recording or otherwise, without the prior permission of the publisher.

Library of Congress Cataloging in Publication Data
Main entry under title:

The Second X and women's health.

(Genes and gender ; 4)
1. Women—Health and hygiene. 2. Sex differences. 3. Women—Mental health. 4. Women—Psychology. 5. Human genetics. 6. Sexism in medicine. I. Fooden, Myra. II. Gordon, Susan. III. Hughley, Betty. IV. Series.
BF341.G39 no. 4   155.2'34s [613'.0424]   83-5552
[RA564.85]
ISBN 0-87752-223-5

CONTENTS

	Page
Genes and Gender Update	
Ethel Tobach	7
Update on Genetics	
Victoria H. Freedman	29
What's New in Endocrinology? Target: Sex Harmones	
Susan Gordon	39
Sex and Temperament Revisited	
Jagna Wojcicka Sharff	49
The Myth of Assembly-Line Hysteria	
Ben Harris	65
Genes and Gender in the Workplace	
Judith S. Bellin and Reva Rubenstein	87
Reproductive Hazards in the Workplace: A Course Curriculum Guide	
Wendy Chavkin, Ruthann Evanoff, and Ilene Winkler with Ginny Reath	101
Sexism in Gynecologic Practices	
Marji Gold	133
The Effects of Childbearing on Women's Mental Health: A Critical Review of the Literature	
Susan Bram	143
Women's Mental Health	
Myra Fooden	161
The Health of Older Women in Our Society	
Georgine M. Vroman	185
Sexism and Racism in Health Policy	
June Jackson Christmas	205
Epilogue	
Betty Hughley and Betty Rosoff	217
Some Additional Resources for Material on Women's Health	221
Biographical Sketches	223

GENES AND GENDER UPDATE

Ethel Tobach, Ph.D.

Periodically the Genes and Gender Collective examines the extent to which the issues of "genetic determinism" remain a widely accepted ideology requiring exposure as a perversion of science and an instrument of oppression of the most vulnerable sectors in the U.S.A.: the non-white, those of non-European origin, children and women. When these conferences were started, it was not only to consider "genetic determinism" generally, but it was in response, as well, to its prevalence as an implicitly or explicitly accepted belief held even by women and men who may believe they are in favor of equal rights and human dignity. (Broverman *et al.*, 1981; Coll *et al.*, 1981; Hrdy, 1981; McClintock, 1981; Hines, 1982; Meece *et al.*, 1982; for an historical example see Mitchell, 1980.) Unfortunately, genetic determinism still prevails as the dominant ideology in science (Davis, 1975, 1980; Altman, 1976; Levi, 1979; Urdy, 1978; Kolata, 1979; Wilson, 1979; DeCatanzaro, 1981; Letters to Science, 1981; Naftolin and Butz, 1981; Baer and McEachron, 1982; McEachron and Baer, 1982; Slade and Hoffman, 1982) and in U.S. Society in general (for a good review of this see Goodman, 1982). "Genetic determinism" is derived from the idea that genes and genetic processes are the final explanation for all physiological and behavioral differences among individuals and groups; specifically they are considered to be responsible for differences in their abilities, capacities and achievements.

The word "final" is very important here. It is generally agreed that genetic processes are present in all living organisms, and that they are functions of specialized biochemical configurations. However, they function *only* on the biochemical level. Many other life-promoting and sustaining processes, such as the integration of external changes (e.g., in the physical and social environment) with internal changes (e.g., ingested food), by definition, include the biochemical functions which are expressions of genes. The genetic determinist, however, sees these other processes as significant only in that they modify, to a limited extent, the ways in which the genes function; for the genetic determinist, the genes are most significant in the "final" determination of how all the other processes will bring about the characteristics of the organism.

The questions posed by the genetic determinists are: is the genetic material that was present in the egg and the sperm that one gets from one's mother and father, respectively, the most significant aspect of one's life? Do these genes "program" one's life? Do they determine what one will be, how one differs from others?

The answer lies in understanding what genetic material is. Certainly the "gene" is a contributing factor to one's life. But what it contributes is of a special character: its most direct contribution is biochemical and only biochemical. The kinds of biochemicals the "gene" produces, or the kinds that it does not produce, are directly linked up, not only to *how* the gene is built, (that is, the specific DNA make-up), but also to the biochemical setting in which it finds itself. The biochemical and physical setting of a particular gene is the product of the total experience of the organism in which it is located: the food ingested; the chemicals, or forms of radiant energy acting on the organism, and the products of other genes in the organism. One knows which gene (biochemical structure) is present, and how it is built and functions, by studying the changes in the organism's structure and function brought about by changing the conditions in which the gene functions. (Waddington, 1953; Wallace, 1979.)

For example, a gene may have a structure which prevents the individual from making use of certain foods in its diet, because the proper enzyme is missing. This is the case with phenylketonuric babies (PKU). A necessary biochemical process cannot take place and the baby does not develop "normally"; when the diet of that baby is properly controlled, the baby is normal.

Some characteristics clearly visible to the observer, like colors of eyes, hair, skin, etc., and skeletal structures, like shapes of head and noses seem to be "determined" by genes. But even these are the result of all that happened to every aspect of the individual: where and how it was fed and taken care of, what other physical and biochemical factors were introduced at what time of its life, as well as genetic functions, etc. For example, if a baby was brought up in a society in which it was thought a good idea to bind the head in a certain way, the shape of its head would be the final integration of all the changes during its whole life history: while it was in the womb, what it ate, and how the head was bound, the stage of development at which it was done, etc. There is no single gene for bones that turn out long or short; for a strong heart or a weak heart. There are genes which make certain biochemicals, which, if given the right circumstances, can play a role in producing long bones or short bones (for example, some individuals are stunted in growth because of poor diet).

The acceptance of genetic determinism by those who believe they are actively fighting sexism is particularly serious. Genetic determinism is expressed variously. Some believe that there are certain "biological" limitations and characteristics which are responsible for the ways in which women as persons deal with the fact that they are women (Henley, 1977). "Inborn" hormonal processes, neurophysiological characteristics and anatomy are thought to be responsible for the way women "handle" reproduction (having babies), respond to men and social situations, see and cope with the world in its physical aspects, or the way they think (Lewis, 1978). To some, these differences seem to be the essence of "womanhood" and are to be cherished or at least acknowledged (for an example of the effect of such theories, see Chernin, 1981). Their position is that if these differences are accepted, women will no longer fight losing battles but rather see these characteristics as strengths in the struggle for a life with dignity.

The Genes and Gender Collective attempts to call attention to this kind of thinking, analyze its scientific merit, and demonstrate how it affects the ways in which people are challenged in their demands for equality of training, jobs and pay. A review of events in our country testifies to the divisiveness of racism and sexism which leads to differentials in pay for work (Anderson, 1981; Anonymous, 1982; Reese, 1981; Russo et al., 1981; Porter, 1982). Those who invest in industry benefit from the resulting lower costs of production and of the maintenance of life systems at work places and elsewhere.

Discriminatory standards result as well for hospitalization (Shuval et al., 1967; Singer, 1967; Gayles, 1974; Bell and Mehta, 1980; Adebimpe, 1981), medical treatment (Bagley, 1983; Bart, 1973; Wilkinson, 1974; Jackson, 1976; Armitage et al., 1979; Bowen, 1979; Verbrugge, 1980; Rheinhold, 1981) and education (Kutner and Brogan, 1970; Schumer, 1971; Clark, 1980; Leacock and Helmrich, 1982); sexism and racism lead to cost-benefit ratios which mean higher investment dividends (Burrows, 1980). It is not surprising that the concept of genetic determinism continues to be the dominant theoretical position in every aspect of academic, scientific and professional life. Once its repressive operation in daily life is understood, the possibility of its defeat in the scientific community is strengthened. And so the need for ongoing study and discussion of this "anti-human ideology" calls for conferences in which the question of its effects is raised and answered.

Genetic determinism is an old concept and has appeared in many forms: social Darwinism (Tobach, 1974, 1976); Freudian psychology, Skinnerian psychology and ethology (Tobach, 1982). The most recent expression of genetic determinism is the subdiscipline of biology called "sociobiology." (See Vol. II, this series) This approach to the evolution and development of social behavior is not new. (Hamilton, 1964) It is "new" only in the sense that it has received its most popularized treatment recently in the writings of experts in biology: Wilson (1975); Dawkins (1976); Wallace (1979) and others, and consequently became established in academic courses of study.

The sociobiological explanation of human behavior rests on the concept that every organism is simply a vehicle by which the genes perpetuate themselves in future generations. Therefore, every individual will do only that which is of benefit to its own reproduction, and to assure that its offspring will survive and be able to continue its genes in future generations. At face-value, this "genetic force" would seem to lead to constant conflict among all individuals and result in little behavior of a cooperative, giving, or "altruistic" nature. However, as Darwin and others after him pointed out, there are many instances in which individuals seem to cooperate with each other, even to the detriment of their own survival. This is seen in parental behavior; in group behavior in which some members of the group fall ready prey to those which are hunting them, while the rest of the group escapes; and certainly in the human species where all cultures extol the virtues of some form of altruistic behavior.

It should be noted that in the dictionary, altruism is defined in the context of *human,* societal, behavior: "unselfish concern for the welfare of others"; opposed to egoism (which is defined as "the tendency to be self-centered, or to consider only oneself and one's own interests; selfishness"). (Webster's, 1966) This definition of altruism emphasizes that the individual who performs the giving act is not going to benefit in any way from the act. In sociobiology, however, any altruistic action is a behavior designed to ensure the perpetuation of the individual's genes in future generations; in other words, the action is based on the selfishness of the actor.

According to the sociobiological way of thinking, the formation of any kind of bond between a sperm/egg-donor and a sperm/egg-recipient comes about because each needs the other in order to pass on its genes. Parental behavior is similarly based: an investment must be made which ensures that the young survive to perpetuate the parents' genes. The amount of investment will determine the kind of parental bond and behavior seen. For example, sociobiologists stress the difference between the parent which incubates and nurtures the young and the parent which spends relatively little time with the offspring after fertilization of the egg has taken place. The incubating parent (which may be either the sperm-donor, as in the case of sea-horses, or egg-donor/sperm-receiver, as in cats) must spend time and energy with the offspring. The offspring incubator and nourisher is seen as specially selected in evolution for characteristics which will increase its chances for doing a good job with the offspring.

Thus, for the sociobiologist, when the incubating animal is female (or a woman) she will be characterized by behavior patterns which will entice the most efficient male (or man) to provide food and shelter while she is incubating the young whether they be young which are developing external to the body, or whether they are developing inside the body of the individual which provided a place for fertilization to take place. (It should be noted that we are using this type of language because there are animals in which it is the sperm donor which is the incubator, e.g., amphibians; fish; birds.)

The sociobiological concept of the gene extends to all aspects of behavior. Thus, besides the social behavior which makes for reproductive efficiency, the degree to which an individual will show cooperative behavior to another is dependent on the number of genes they hold in common. Offspring (or children) which have the same amount of genes from each of two parents have more genes in common with each other than with either of their parents. Therefore, according to sociobiologists, there is more likely to be cooperation between siblings than between a parent and an offspring (for example, social insects). Cousins will be cooperative depending on how closely related they are to each other. Members of the same community will be more likely to share more genes with each other than with other individuals from other communities. According to sociobiologists, therefore, in the early evolution of human beings, individuals with closely shared genes lived in small groups which made survival possible (Wallace, 1979). Associated with this idea is the concept that, because their genes are different, communities are suspicious of, and consider themselves better than, individuals from other communities. This ensures the cohesiveness of each group in defending itself and leads to aggression towards others in order to survive. In this way, sociobiology teaches that genes and the number of shared genes explain interactions among individuals of different gender and group history. Differences in physiological morphological characteristics are viewed as strategies designed in evolution to ensure survival of the species. In other words, racism and sexism which are "justified" by genetic determinism, are further "supported" by sociobiology which claims to demonstrate how they evolve through a natural selection process.

Is Wilson aware of the political significance of what he says? In an interview by a Japanese science reporter in 1981, he considers himself "apolitical," (Miura, 1981) although he is clearly knowledgeable about current views of the political implications of his theoretical approach (Wilson, 1977). For example, in discussing human nature, Wilson tells us that we cannot ignore the struggle of women for equal rights (Wilson, 1978). But, since all behavior is determined by the genes one has he sees a real problem in dealing with the differences between women and men—differences which are predetermined and which without genetic engineering cannot be controlled.

He gives society three choices.

> (1) *"Condition its members so as to exaggerate sexual differences in behavior.*
> This is the pattern in almost all cultures. It results more often than not in domination of women by men and exclusion of women from many professions and activities. But . . . In theory . . . at least, a carefully designed society with strong sexual divisions . . . could be richer in spirit, . . . and even more productive than a unisex society. Such a society might safeguard human rights even while channeling men and women into different occupations. Still some amount of social injustice would be inevitable, and it could easily expand to disastrous proportions." (Pp. 132–133)
> (2) *"Train its members so as to eliminate all sexual differences in behavior.*
> "It could result in a much more harmonious and productive society. Yet, the amount of regulation required would certainly place some personal freedoms in jeopardy, and at least a few individuals would not be allowed to reach their full potential." (P. 133)
> (3) *"Provide equal opportunities and access but take no further action.*
> "Even with identical education for men and women and equal access to all professions, men are likely to maintain disproportionate representation in political life, business and science. Many would fail to participate fully in the equally important, formative aspects of child rearing." (P. 133)

The words may change, but the melody lingers on. The genetic destiny of women must be accepted *or else social injustice,* that is, limitation of personal freedom, *will prevail.* He further stresses that no egalitarian society has been able to, or will be able to, countermand the "biological determination" of women's gender role.

Another example of his writing which, according to him, should be considered "apolitical" is his discussion of genetic engineering.

According to Wilson, humanity has been disillusioned in the myths of religion and Marxism which tried to explain the complexity of life. He proposes a new myth, the evolutionary epic, which is based on his view of genetics. The myth of the evolutionary epic which explains all is based on the possibility of genetic manipulation. "Jehova's challenges have been met and scientists have pressed on to uncover and to solve even greater puzzles. The physical basis of life is known; we understand approximately how and when it started on earth. New species have been created in the laboratory and evolution has been traced at the molecular level. Genes can be spliced from one kind of organism to another." (Wilson, 1978; P. 198). Thus, there is hope for a "democratically contrived eugenics" which will solve the dilemmas he sees confronting humanity.

Wilson is aware of the unpopularity of relating these advances in genetic manipulation to the genetic control of behavior. In an article in *Daedalus* (1977) he acknowledges the complicated, detailed procedures that geneticists must use to tease out the relations between the biochemical level of the gene and any structural or functional trait. Nonetheless, he says, "Human social behavior can be evaluated in the same way." (Wilson, 1977, P. 131). He concludes his excursion into genetic determinism with the following paragraph:

> "My overall impression of the available information is that *Homo sapiens* is a typically animal species with reference to the quality and magnitude of the genetic diversity affecting behavior. If the comparison is correct, the psychic unity of humankind has been reduced in status from a dogma to a testable hypothesis. This is not an easy thing to say in the present political ambiance of the United States, and it is regarded as punishable heresay in some sectors of the academic community. But the idea needs to be faced squarely if the social sciences are to be entirely honest. I cannot regard it as dangerous. Quite the contrary: the political consequences of its objective examination will be determined by our value system, not the reverse. It will be better for scientists to study the subject of genetic behavioral diversity than to maintain a conspiracy of silence out

of good intentions and thereby default to ideologues." (Wilson, 1977, P. 133).

Wilson's position, however, is that contemporary value systems should be replaced by an appreciation of the evolutionary epic myth: the myth that genes determine the nature of the human mind, human morality, and the human propensity for philosophizing. All that is needed to carry out this replacement is knowledge of the appropriate set of multiple genes to be manipulated. "For example, diabetes and schizophrenia possess moderate genetic components. The multiple genes underlying them produce a stronger tendency to develop the traits; they also prescribe the range of possible manifestations that are probable under specific environmental conditions. In a parallel way basic human social behavior . . . emerge as outermost phenotypes following behavioral development . . . constrained by the interaction of polygenes with the environment. With reference to this interaction, there is no reason to regard most forms of human social behavior as qualitatively different from physiological and non-social psychological traits." (Wilson, 1977, P. 132)

Wilson, an accomplished debater, anticipated the obvious response to his theory of human nature: if some humans have already evolved the capacity to consider "democratically contrived eugenics" a possible solution to their problems, there are others who have not reached that conclusion. However, he proposes that the basis for the inability of people to see this kind of eugenics in a positive light is related to the peculiarly human conscience which has been inherited as part of the human nervous system.

In other words, inherited human characteristics made it possible for the species to develop a technology for genetic engineering. Through genetic engineering, people would control genes for social behavior and could evolve patterns for living in hive-like societies like ants or bees, or could be programmed to have a father, a mother and a requisite number of children in its social or family unit like some apes. To do this, people have to know which genes are responsible for different kinds of social behavior.

Wilson proposes that the amassing of this type of information is stymied by the conscience encoded in the nervous system. Somehow, in addition to inheriting the ability to develop the technique, the human species has inherited a belief system which goes against programming individuals through the technique of genetic engineering. He is proposing, of course, an up-dated version of eugenics. The eugenics movement was based on another pseudoscience which originated after Darwin's theory of evolution through the processes of heredity and natural selection became widely known. It was a movement to help evolution along by legally determining who could have children and who could not in order to "improve" the species. It was the basis in the United States for formulating immigration policy and the treatment of Blacks, Amer-indians and the poor, and by the Nazis in Germany (Chase, 1977; Bennett, 1974).

Wilson states that the dilemma brought about by technology and conscience will be resolved sometime in the future. Robert A. Wallace, however, another sociobiologist, is ready to apply sociobiological principles to contemporary social practices. He wishes to use the principles of "shared genes" to combat discrimination, which he says he opposes. He makes the following suggestions:

> "For example, we might begin by discouraging pride in our ancestry. Forget 'White supremacy' and 'Black is beautiful'. If blacks and whites were to give up their racial pride to freely intermingle and intermarry, we would destroy one means of discrimination. Do not be proud of your Italian ancestry. If Italian Americans deny their heritage and move out of their neighborhoods to disperse into the population at large, their descendants will be unnoticed, never worrying about being defamed. If Jews relinquish their ethnicity and move to Arkansas, their descendants will go unnoticed." (Wallace, 1979, Pp. 206–207)

It should be noted that on the dust jacket of this book, Wilson says that the book is a "soundly argued and written popular sermon on the biological realities of the human condition."

The system of sociobiological ideology is internally consistent. For example, Wilson's sociobiology explains the changes from a slavery-founded social system to a non-slave owning system as dependent on a preprogrammed biological reason for such a change. He uses a quotation from Lionel Trilling's book "*Beyond Culture*" to explain slavery and the change from slavery: "there is a hard, irreducible, stubborn core of biological urgency, and biological necessity, and biological *reason*, that culture cannot reach and that reserves the right, which sooner or later it will exercise to judge the culture and resist and review it." (Wilson, 1978, P. 80). It is reasonable, then, that this same preprogrammed biological necessity will keep cultural evolution on the correct biological tracks in regard to social systems which generate weak cultures.

> ". . . biological evolution is always quickly outrun by cultural change. Yet the divergence cannot become too great, because ultimately the social environment created by cultural evolution will be tracked by biological natural selection. Individuals whose behavior has become suicidal or destructive to their families will leave fewer genes than those genetically less prone to such behavior. Societies that decline because of a genetic propensity of its members to generate competitively weaker cultures will be replaced by those more appropriately endowed. I do not for a moment ascribe the relative performance of modern societies to genetic differences, but the point must be made: there is a limit, perhaps closer to the practices of contemporary societies than we have had the wit to grasp, beyond which biological evolution will begin to pull cultural evolution back to itself." (Wilson, 1978, P. 79–80).

In a later book, *Genes, Mind* and *Culture* (Lumsden and Wilson, 1981), he repeats this concept. Through mathematical models developed with his co-author, he proposes that there are epigenetic rules which govern human behavior. "Epigenetic rules are ultimately genetic in basis" (P. 370) although epigenesis is "the process of interaction between genes and the environment." (P. 370) In the last chapter of the book, he once again gives an application of sociobiological principles:

"A society that chooses to ignore the implications of the innate epigenetic rules will still navigate by them and at each moment of decision yield to their dictates by default. Economic policy, moral tenets, the practices of child rearing, and virtually every other social activity will continue to be guided by inner feelings whose origins are not examined. Such a society must consult but cannot effectively challenge the oracle residing within the epigenetic rules. It will continue to live by the "conscience" of its members and by "God's will." Such an archaic procedure . . . will perpetuate conflict and relentlessly drag humanity along what is at best a tortuous and dangerous path. On the other hand, the deep scientific study of the epigenetic rules will call the oracle to account and translate its commands into a precise language that can be understood and debated. Societies that know human nature in this way might well be more likely to agree on universal goals within the constraints of that nature. And although they cannot escape the inborn rules of epigenesis, and indeed would attempt to do so at the risk of losing the very essence of humanness, societies can employ knowledge to guide individual behavior and cultural evolution to the ends upon which they agree. . . . We should keep in mind that most of the wonderous inventions of science and technology serve in practice as enabling mechanisms to achieve territorial defense, communications of tribal ritual, sexual bonding, and other ancient sociobiological functions." (Pp. 358–360)

Genetic determinism in its various forms permeates every facet of our lives: in our books (Francis, 1981), movies, theater, magazines (Medical Aspects of Human Sexuality, 1981; Fleming, 1981), newspapers (for an outstanding example of good newspaper coverage see Comer and Poussaint, 1979; also Brody, 1982; Webster, 1982; Zukerman, 1982; Webster, 1982) and television (Steinman, 1982). The ultimate perversion of the long history of human evolution is seen in the sociobiological concept of altruism. They have perverted that human quality which can see beyond the individual's own needs into the ultimate expression of selfishness: the perpetuation of one's own self and kind. Accordingly, the concepts of philanthropy and social justice are seen as nothing but the ultimate attempts of the *species* to preserve itself.

An exercise in the application of sociobiology to contemporary issues in international relationships may be used to demonstrate its "epigenetic rule." The movement against nuclear war would be seen as an example of the attempt by some people to preserve the species in response to the "genetic" drive. At the same time the desire of every nation to have its own stockpile of nuclear weapons is the expression of the distrust of others with different genes. The outcome of this battle between these two great evolutionary forces is dependent on an inherited species conscience encoded in the human nervous system. Presumably, the resolution of this "dilemma" will depend on natural selection by nuclear warfare of the individuals best equipped to survive and perpetuate their genes.

Is there any evidence for accepting the ethological and sociobiological concepts of genes? As will be seen by Friedman's paper in this volume, geneticists can be clear as to how genes contribute to individual differences. The "evidence" for genetic determinism is open to two very basic criticisms: 1. Is the statement that a characteristic is "genetically determined" supported by experimental evidence? In other words, is it possible to trace a specific biochemical structure and pathway of function that bring about that characteristic under known circumstances? For example, knowing that a genetic configuration is responsible for the absence of an enzyme tells us how a gene can bring about a condition called phenylketonuria (PKU), if certain conditions of diet exist. The presence of PKU can be defined very clearly by biochemical analysis. But, how phenylalanine metabolism, which is affected by that gene, makes for differences in mental growth and development is still not understood.

Another example of this problem is Down's Syndrome. Although the chromosomal characteristic of this syndrome has been extensively and intensively studied, the relationships between the differences in chromosomes and the functional characteristics of such individuals are still not known.

The second problem with "evidence" of genetic determinism

stems from the first. Genetic/biochemical facts are only the beginning of the story of how differences come about. Furthermore, saying that something is "genetic" is not only a small part of the problem of understanding differences, but it is trivial because all living things are defined by the fact that they can reproduce themselves, that is, they have genes which can duplicate themselves. This distinguishes living from nonliving matter. To say something is living, is to say that it is "genetic"; otherwise it would not be "living." Nothing is added to our understanding of any of the behavioral characteristics usually discussed by genetic determinists. To understand how humans got to be the way they are, their entire history, including their biochemistry (for example, their genes and their nutrition) and the societal circumstances in which they grew and developed would have to be known. To answer the question of individual differences, scientifically valid methods would have to be used and valid questions which could be answered reliably would have to be asked. This approach to understanding people is difficult and expensive. It would require large investments of time, effort and resources. (An example of this is discussed by Rome, 1982.) It cannot be undertaken when the ideology of the society has other priorities, such as large military budgets.

The ideology of genetic determinism is useful in convincing people about the values they should have. For example, popular "science" writers, say that men are genetically more reliable as militarists than women. The struggle against registration and a military draft is a recent arena in which arguments based on supposed differences between women and men are being used as a diabolically divisive instrument to come between women and men in their critical fight against the nuclear destruction of humanity.

Government supported research and publications perpetuate the idea that gender is determined by hormones, not societal processes (Anonymous, 1982). Medical researchers in the Dominican Republic found a population of people with a biochemical (genetic) error which delays development and which produces sterility. A research report by Sara M. Vélez (1977) showed the importance of societal values in determining whether these people were brought up in either a feminine or masculine role and whether they continued in this role after the physiological and structural changes occurred. And yet, the research with, and medical treatment of, the people in this population are focussed only on the hormonal and biochemical processes involved. (Anonymous, 1982).

In educational practices, genetic determinism remains the accepted dogma. For example, Jensen's book explaining how to pervert statistics to justify the racism in our educational system is widely hailed by academics (Leacock and Helmreich, 1982). Women are considered unable to be good in mathematics because their brains are not "built right." (for views of this concept, see Meece *et al.*, 1972; Letters to *Science*, 1981). In reviewing the scientific literature, in surveying the college and high school catalogs which list courses in sociobiology and ethology as *bona fide* scientific disciplines, and participating in the continuous struggles for equality of women and minorities, it becomes clear that far from being exposed and defeated, the dogma of genetic determinism is stronger than ever.

It must be recognized that in the daily struggle to maintain oneself, the scientific argument about genetic determinism seems very remote. Loss of jobs, loss of child welfare support, poor pay and poor working conditions need to be fought in the societal arena where people of good will can come together in common struggle. (Leacock, 1981) However, when people are schooled and trained from earliest childhood to believe that they are handicapped by some mysterious characteristics with which they were born; when they see this idea mirrored in every aspect of their cultural and personal life; when these so-called unchangeable differences are used as the explanations of their supposed personal and group failures in life; to gather up the courage and stamina to fight for equality and justice is made more difficult. It is a particularly human characteristic to be able to plan the future and understand the consequences of one's acts. This characteristic was probably derived from activities in the earliest stages of human evolution which controlled the environment in ways which other species did not. These activities of control and planning were powerfully integrated factors in the development of human structure, physiology and behavior. It is critical that people today be firm in their conviction that they can control the future and solve the problems which seem to be putting that future in doubt.

ACKNOWLEDGEMENTS

I wish to thank Janet Ploss, Betty Rosoff, Georgine Vroman and Myra Fooden who critically reviewed the manuscript. Of course, they are in no way responsible for its errors and failures.

REFERENCES

Adebimpe, V. R.
   1981   Overview: White Norms and Psychiatric Diagnosis of Black Patients. American Journal of Psychiatry, 138, 279–285.

Altman, L. K.
   1976   Doctor Regrets Minorities View. New York Times, May 22, P. 17. (Article about Bernard D. Davis; see below).

Anderson, M.
   1981   Neither Jobs nor Security. Employment Research Associates, 400 South Washington Avenue, Lansing, Michigan, 48933.

Anonymous
   1982   Data Show Women's Pay Still Lags Behind Men's. New York Times, March 7, P. 25.

Anonymous
   1982   Pseudohermaphrodites: Models of the Developing Male. Research Resources Reporter, 6 (5). U.S. Department of Health and Human Services, Bethesda, Maryland.

Armitage, K. J., L. J. Schneiderman, and R. A. Bass
   1979   Response of Physicians to Medical Complaints in Men and Women. Journal of the American Medical Association, 24, 2186–2187.

Baer, D. and D. L. McEachron
   1982   A Review of Selected Sociobiological Principles: Application to Hominid Evolution. I. The Development of Group Social Structure, Journal of Social Biological Structure, 5, 69–90.

Bagley, C.
   1973   Occupational Class and Symptoms of Depression. Social Science and Medicine, 7, 327–340.

Bell, C. C. and H. Mehta
   1980   Misdiagnosis of Black Patients with Manic Depressive Illness: II Journal of the National Medical Association, 72, 141–145.

Bennett, A. P.
   1974   Eugenics as a Vital Part of Institutionalized Racism. Freedomways, 14, 111–126.

Bowen, M. E.
   1979   *Williams Obstetrics* on Abortion. Man and Medicine, 4, 205–232.

Brody, J. E.
   1982   Female Hyenas Reign. New York Times, August 31, P. 1 and 4.

Broverman, D. M., W. Vogel, E. L. Klaiber, D. Majcher and D. Shea
   1981   Changes in Cognitive Task Performance Across the Menstrual Cycle. Journal of Comparative and Physiological Psychology, 95, 646–654.

Burrows, V. (Editor)
1980 Racial Discrimination and its Effects on the Economic Rights of US Women. United States Seminar Sponsored by the Women's International Democratic Federation (WIDF) and Women for Racial and Economic Equality (WREE), 6/6–8/1980; New York.

Chase, A.
1977 The Legacy of Malthus: The Social Cost of the New Scientific Racism. Knopf, New York.

Chernin, K.
1981 How Women's Diets Reflect Fear of Power. The Sunday New York Times Magazine, 10/11/81, P. 38–50.

Clark, T. B.
1980 Campuses and the Feds: Dancing to Washington's Regulatory Tune. National Journal, 11/29/80, P. 2016–2024.

Coll, C. G., Sepkoski, C. and B. M. Lester
1981 Cultural and Biomedical Correlates of Neonatal Behavior. Developmental Psychobiology, 14, 147–154.

Comer, J. P. and A. P. Poussaint
1979 Children Aren't Predestined by 'Bad Seed'. Sunday Sun-Times (Boston), 11/4/79, P. 18.

Davis, B. D.
1975 Social Determinism and Behavioral Genetics. Science, 189, 1049.

———
1980 Pythagoras, Genetics and Workers' Rights. New York Times, August 14, 1980, P. 23. (See L. K. Altman above)

Dawkins, R.
1976 The Selfish Gene. Oxford University Press, New York.

DeCatanzaro, D.
1981 Suicide and Self-Damaging Behavior: A Sociobiological Perspective. Academic Press, New York.

Fleming, A. T.
1981 Women and the Spoils of Success. Sunday New York Times Magazine, August 2, P. 30–31.

Francis, D.
1981 Reflex. Fawcett-Cress, New York (P. 189).

Gayles, J. N., Jr.
1974 Health Brutality and the Black Life Crisis. The Black Scholar, 5, 2–9.

Goodman, E.
1981 Latest Debate Over 'Designer Genes' Cut Out of Old Cloth. Boston Globe, December 23, 1980, P. 27.

Hamilton, W. D.
1964 The Genetical Evolution of Social Behavior. I and II. Journal of Theoretical Biology, 7, 1–52.

Henley, N. M.
1977 Body Politics: Power, Sex, and Nonverbal Communications. (Pat-

terns of Social Behavior series.) Englewood Cliffs, New Jersey, Prentice-Hall.

Hines, M.
1982 Prenatal Gonadal Hormones and Sex Differences in Human Behavior. Psychological Bulletin, 92, 56–80.

Hrdy, S. B.
1981 The Women that Never Evolved. Harvard University Press, Cambridge, Mass.

Jackson, J. J.
1976 The Plight of Older Black Women in the U.S. The Black Scholar, 7, 47–55.

Kolata, G. B.
1979 Sex Hormones and Brain Development. Science, 205, 985–987.

Kutner, N. G. and D. R. Brogan
1980 The Decision to Enter Medicine: Motivations, Social Support, and Discouragements for Women. Psychology of Women Quarterly, 5, 341–357.

Leacock, E.
1981 History, Development, and the Division of Labor by Sex: Implications for Organization. Signs: Journal of Women in Culture and Society, 7, 474–491.

———— and W. B. Helmreich (Eds.)
1982 IQ: Science or Politics? A Symposium. Journal of Academic Skills, 3, 1–62.

Letters to Science
1981 Letters in response to an article on: Mathematical Ability: Is Sex a Factor? Science, 212, 114–121.

Levi, L.
1979 Psychosocial Factors in Preventive Medicine. In Healthy People, The Surgeon General's Report on Health Promotion and Disease Prevention, U.S. Department of Health, Education and Welfare, U.S. Government Printing Office, Washington, D.C.

Lewis, H. B.
1978 Psychology and Gender. In Genes and Gender: On Hereditarianism and Gender. Gordian Press, New York.

Lumsden, C. J. and E. O. Wilson
1981 Genes, Mind and Culture. Harvard University Press, Cambridge, Mass.

McEachron, D. L. and D. Baer
1982 A Review of Selected Sociobiological Principals: Application to Hominid Evolution. II. The Effects of Intergroup Conflict. Journal of Social Biology Structure, 5, 121–139.

McClintock, M. K.
1981 Social Control of the Ovarian Cycle and the Function of Estrous Synchrony. American Zoologist, 21, no. 1, 246–254.

Meece, J. L., J. E. Parsons, C. N. Kaczala, S. B. Goff and R. Putterman

    1982   Sex Differences in Math Achievement: Toward a Model of Academic Choice. Psychological Bulletin, 91, 324–348.

Mitchell, R.
    1980   Euthenics at Vassar College. Paper presented at the Eastern Psychological Association Convention.

Miura, K.
    1981   Sociobiology Clarifies Human Nature. The Kagazku Asha, 41, February, 34–38. Translated by Shinobu Kitayama.

Naftolin, F. and E. Butz (eds.)
    1981   Sexual Dimorphism. Science, 211, 1263–1324.

Porter, S.
    1982   The Budget Cuts and Older Women. The New York Daily News, October 4, P. 35.

Reese, E. P.
    1981   The Status of Women in Psychology. Report for the Committee for Women in Psychology, American Psychological Association, August 27, 1981.

Rheinhold, R.
    1981   Care of Newborns Improves, But High-Risk Births Persist. The New York Times, December 28, 1980, P. 9.

Rome, E. R.
    1982   Letter to the Editor, The New York Times, August 26, 1982, P. 30. Premenstrual Tension.

Russo, N. F., E. L. Olmedo, J. Stapp, and R. Fulcher
    1981   Women and Minorities in Psychology. American Psychologist, 36, 1315–1363.

Schumer, F. R.
    1981   A Question of Sex Bias at Harvard. Sunday New York Times Magazine, October 18, 1981.

Scully, D. and P. Bart
    1973   A Funny Thing Happened to Me on the Way to the Orifice: Women in Gynecology Textbooks. American Journal of Sociology, 78, 1045–1052.

Shuval, J. T., A. Antonovsky, and A. M. Davies
    1967   The Doctor-Patient Relationship in an Ethically Heterogeneous Society, Social Science and Medicine, 1, 141–154.

Singer, D. D.
    1967   Some Implications of Differential Psychiatric Treatment of Negro and White Patients. Social Science and Medicine, 1, 77–83.

Slade, M. and E. Hoffman
    1982   Biology and Destiny. New York Times, 5/23/82, P. 14.

Steinman, S. and A. Kornhaber
    1982   Talk Program "Human Behavior" on Grandparents. Channel CNN, 9/4/82.

Tobach, E.
    1973   Social Darwinism Rides Again. In The Four Horsemen: Racism,

Sexism, Militarism and Social Darwinism. New York, Behavioral Publications, 1973, Pp. 97–123.

1976 Behavioral Science and Genetic Destiny: Implications for Education, Therapy, and Behavioral Research. In Genetic Destiny: Scientific Controversy and Social Conflict. E. Tobach and H. M. Proshansky (eds.), New York, AMS Press, Pp. 152–158.

1982 The Synthetic Theory of Evolution. In Evolution and Determination of Animal and Human Behaviour, G. Tembrock and H.-D. Schmidt (eds.), Vep Deutscher Verlag des Wissenschaften, Berlin, Pp. 27–39.

Urdy, J. R.
1978 Differential Fertility by Intelligence: The Role of Birth Planning. Social Biology, 25, 10–14.

Velez, S. M.
1977 Gender Role of Male Pseudohermaphrodites in the Dominican Republic. Master of Arts (Anthropology) Dissertation, Hunter College, The City University of New York.

Verbrugge, L. M.
1980 Sex Differences in Complaints and Diagnosis. Journal of Behavioral Medicine, 3, 327–356.

Waddington, C. H.
1953 Genetic Assimilation of an Acquired Character. Evolution, 72, 118–126.

Wallace, B.
1966 Chromosomes, Giant Molecules and Evolution. New York, Norton.

Wallace, R. A.
1979 The Genesis Factor. Morrow, New York.

Webster, B.
1982 Infanticide: Animal Behavior Scrutinized for Clues to Humans. New York Times, 8/17/82, C1–4.

Webster's New World Dictionary
1966 College Edition, New York.

Wilkinson, D. Y.
1974 For Whose Benefit: Politics and Sickle Cell. The Black Scholar, 5, 26–31.

Wilson, E. O.
1975 Sociobiology. Belknap Press, Cambridge, Mass.

1977 Biology and the Social Sciences. Daedalus, 106, 127–140.

1978 On Human Nature. Harvard University Press, Cambridge, Mass.

Wilson, G.
1979 The Sociobiology of Sex Differences. Bulletin of the British Psychological Society, 32, 350–353.

Zukerman, E.
1981 Why More Women Aren't Superstars in Music. The Sunday New York Times February 22, 1981, Section II, P. 1 and 19.

UPDATE ON GENETICS

Victoria H. Freedman, Ph.D.

The concept of the gene as the unit of heredity developed as the mysteries of heredity were unraveled. The first studies were carried out at the beginning of this century by the Austrian monk, Gregor Mendel, who studied the patterns of inheritance in pea plants. However, the basic mechanisms of heredity were elucidated by careful analyses of the genetics of simple microorganisms, such as bacteria and fungi. (Reviewed by Prober and Ehrman in Genes and Gender I 1978). These organisms were used for study because they grew easily in the laboratory, produced a new generation in a matter of 20 minutes or so, and could be grown in vast quantities, thereby facilitating biochemical analyses. The pioneering studies carried out with these microorganisms led to the fundamental idea that each gene gave rise to a specific product (protein). Only when that gene was changed in some way, as might occur when the organism was exposed to a mutagenic agent such as X-rays, was the gene product changed. Thus, each trait of the bacterial cell was considered to be *determined* by the action of a particular gene, and this gene was seen as a single unit, whose sole function was the transmission of genetic information specific for a particular characteristic, or trait.

It is this concept of a single gene controlling a single trait that has remained in our minds. When most of us think of genetics, we think of individuals as merely collections of genes, which in turn give rise to collections of traits, and these collections of traits characterize each of us.

This popular notion of the gene, and of heredity, is to a significant extent, erroneous. First, it is based on the extrapolation of data from the bacterial system to the human system. Of course, the human system is extremely complex, and humans are not just large bacteria. Also, even in bacteria it is now known that bacterial traits are not necessarily the result of the action of single genes. The view of the single gene determining a single trait is too simplistic. Actually, the evidence accumulating now suggests that the gene is just the beginning of a long pathway leading to the final expression of a particular phenotype (or trait), and there are many sites along this pathway where changes (or modulations) may occur.

Furthermore, multicellular organisms such as we are, result from carefully regulated and extremely complex interactions which occur among the products of many genes, and which are, in turn influenced to a greater or lesser degree, by the environment in which that gene product is being expressed. Therefore, the phenotype of any individual cell or individual organism is not fixed, but is subject to changes induced by the environment in which the genes are acting.

There are therefore three fundamental, and new, concepts of the gene which are important in understanding the processes of heredity. First, in some cases, one gene does not determine one protein, but only a part of the protein. For the complete functional protein to be made, several genes must act together. One of the most notable examples are the immunoglobulin genes, which are responsible for the remarkable specificity and diversity of the antibodies made by our bodies to defend us against bacteria, proteins or other organochemicals, that are foreign to the body.

Secondly, the gene is just the beginning of a complex system of molecular activities which comprise the pathway of gene expression. In many cases, the product of that gene has to undergo many changes before it is in the operational, or active form. This is especially true of hormones, which must be in one form to be secreted from the cell which makes them, and in another form to be active.

Thirdly, the cellular environment plays a tremendous role in the final expression of gene products. In one series of experiments, it has been shown that the malignant nature of a mouse tumor cell can be completely reversed when it is experimentally placed in a normal embryonic milieu. That is, a cell from one very special type of mouse tumor, the teratocarcinoma, can function like a normal embryonic cell when surrounded by normal embryonic cells in a developing mouse embryo.

Let us discuss the first case which challenges the popular notion, that one gene gives rise to one particular gene product. While this may actually be the case in some instances, recent advances in molecular biology suggest that it is certainly not quite this simple in many other instances, especially in higher organisms such as mice and people.

In the last several years, a great deal of evidence has accumulated from the study of many organisms showing that most genes (which are composed of DNA) occur in pieces rather than as continuous stretches of DNA. The old concept of "one gene-one enzyme," that is, that each gene by itself codes for a complete product, has therefore undergone a dramatic revision. One of the most exciting discoveries in this field has been in regard to the production of antibodies by the cells that make them. It is now clear that each antibody molecule is coded for by separate genes which combine to give the complete genetic information for the complete molecule. In antibody formation, each gene by itself is not sufficient to code for the entire functional protein (antibody).

In order to understand the genetics of this particular situation, we must first know something of the basic structure of an immunoglobulin (antibody) molecule. Each immunoglobulin molecule consists of two protein chains: a "light" chain and a "heavy" chain. Each light chain, which is a protein chain of about 200 amino acids (the building blocks of proteins), can be subdivided into 2 distinct regions: the variable region, which is different in each light chain molecule, and the constant region, which remains the same in all light chain molecules of a given type in a given individual.

Recently, it has been shown by Brack *et al.* (1978) and Max *et al.* (1979) among others, that most of the variable region is coded for by one gene, called the V gene, and the constant region is coded for by another gene, called the C gene. The remainder of the variable region of the immunoglobulin molecule is coded for by still another gene, which is known as J, since it codes the part of the antibody molecule in which the variable region is *joined* to the constant region. This is shown in the simplified diagram that follows.

The heavy chains of the immunoglobulin molecules, although much larger (about 440 animo acids) also have a variable region and a constant region. It has been very recently shown that the heavy chains are also coded for by separate genes. In this case, however, three genes give rise to the heavy

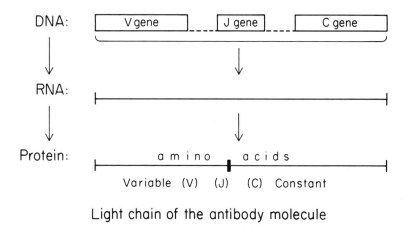

Light chain of the antibody molecule

chain variable region. In addition, there is a J gene and a C gene specific for the heavy chain (Marx, 1981).

The descriptions given above are very simplified, but it is nevertheless clear that the immunoglobulin molecule is a prime example of a situation in which the genetic information of several genes must be combined in order for a functional product to ensue. As researchers continue to probe the genetics of mammalian cells, many more examples of this will become apparent.

The second series of experiments to be discussed suggest that the final phenotype is dependent not only on the synthesis of a particular protein according to the information encoded in the gene for that protein, but also on the ability of the protein to get to another site, either inside the cell or outside the cell where it may carry out its specific function. This is especially important in the case of multicellular organisms, which we are. Thus, a protein may be made by one particular cell and then it may have to get out of that cell, a process known as secretion, in order to be active. In addition, in many cases, the protein is made as part of a much larger protein, which is then sequentially "chopped" down to the right size. The "extra" parts of the protein, appear to function as guides to enable that protein to get through the maze of intracellular organelles to reach the cell surface membrane, the "door" of the cell. At the door, the protein goes through, while the guide (the extra parts of the protein) remain behind.

This kind of scheme has been postulated for the synthesis and secretion of many hormones. One of the most intensively studied is the growth hormone of the anterior pituitary. Blobel and his associates (1980) showed that the cells of the anterior pituitary actually synthesize a slightly larger protein, which is called pre-growth hormone. As the pre-growth hormone is secreted through the membrane of the cell, the extra parts are cleaved off, so that the authentic, functional growth hormone is what is actually found outside the cell. The extra parts, which remain in the cell, form the "signal" to carry the hormone through the cell membrane. This is believed to be a major mechanism in the secretion of cellular products. Thus, hormone synthesis and secretion present a situation in which the original product of the gene is an inactive protein, which must be modified and transported before it is functional.

The third important concept to be discussed is that the cellular environment is integral to the final expression of gene products. This is most frequently studied in developing embryos. The interaction between cells in the early mammalian embryo is important in the differentiation of the various organ systems. In recent years, a number of investigators have studied these interactions by experimentally manipulating whole embryos of mice. One particular technique was pioneered by scientists such as Gardner and Mintz (1975). The mouse blastocyst is an early embryonic stage, in which the fertilized egg has already divided several times, and formed a hollow ball of cells. As development proceeds, the cells at one end of the ball begin to divide more rapidly. These will eventually form the embryo itself. The microinjection technique allows the experimental placement of other cells or groups of cells into the hollow inside of the developing mouse blastocyst. The blastocyst is then reimplanted into the uterus of a pseudopregnant female, that is, a female mouse whose uterus has been hormonally prepared to allow gestation. In many cases, the injected cell becomes part of the blastocyst and participates in the development of the embryo. A significant number of these reimplanted blastocysts actually proceed to develop into viable mice, and after these mice are born, their tissues are analyzed to determine which came from the original blastocyst and which developed from the cell or cells that were experimentally injected.

The perfection of these experimental techniques allowed scientists to ask some fundamental questions about the developmental pattern each embryo proceeds through. The fertilized cells from which the organism or animal originates, proceeds to divide in an original and stepwise fashion, giving rise to specialized cells and tissues which in turn give rise to the complete individual. The intriguing question that one might ask is how determined and unchanging is this pattern? Is it unidirectional? That is, has an adult differentiated cell lost the ability to differentiate in a different way?

To answer questions like these Mintz and her colleagues (1975) carried out a fascinating series of experiments using the procedures outlined above. However, the question they asked was even more interesting. Can a malignant cell, a cell whose "program" is somehow deranged and can no longer function normally, participate in normal development? In mice, as well as in humans, tumors known as teratocarcinomas may arise. These tumors generally develop in gonadal tissue. In mice, they are malignant tumors whose rapid unchecked growth will lead to the death of the host. However, these tumors frequently give rise to a variety of normal differentiated cell types such as nerve, skin, muscle or cartilage.

In her experiments, Mintz and her colleagues (1975) injected the undifferentiated malignant cells of the teratoma into normal mouse blastocysts and then reimplanted these blastocysts into the uteri of suitably prepared female mice. A large number of these blastocysts proceeded to develop, and the mice subsequently born were studied to determine whether the injected malignant cell had contributed to their development. Through the analyses of coat color, the hemoglobin types, various enzymes and the tissue architecture of many organs, these investigators showed that many normal tissues had arisen from the injected teratoma cell. Thus, the malignant cells of the teratoma, a tumor cell that generally kills its host within 3–4 weeks, was capable, when placed in the environment of a normally developing embryo, to participate in normal development and give rise to normally functioning tissues. In addition, subsequent experiments showed that these mice were also capable of mating and producing normal offspring.

These experiments thus suggest that neoplasia, at least in the case of the murine (mouse) teratoma, did not involve a structural change in the genome, but rather a change in gene expression. The experiments further raise the possibility of the reversal of cellular malignancy in a normal cellular environment, thus giving support to the thesis advanced here, namely that in many cases, the expression of a particular cellular trait or traits (the phenotype) is determined not only by the presence of the gene for that trait, but by the cellular environment that the particular gene is operating in.

The examples cited above are all illustrative of the changing concept of the gene. Actually, the gene is just the beginning of molecular processes leading to the final expression of the trait or phenotype. The gene, the gene product, and finally the cell, are subject to numerous processes which subtly alter or significantly modify the expression of the original information. This, on a molecular level, is what generates the marvelous diversity of all living organisms.

REFERENCES

Brack, C., Hirama, M., Lenhard-Schuller, R., and Tonegawa, S.
   1978   A complete immunoglobulin gene is created by somatic recombination. Cell 15: 1–14 (178).

Lingappa, V. R., and Blobel, G.
   1980   Early events in the biosynthesis of secretory and membrane proteins: the signal hypothesis. In Recent Progress in Hormone Research, N.Y., Academic Press 36: 451–475.

Max, E. E., Seidman, J. G., and Leder, P.
   1979   Sequences of five potential recombination sites encoded close to an immunoglobulin K constant region gene. Proc. Nat. Acad. Sci. 76: 3450–3454.

Marx, J.
   1981   Antibodies: getting their genes together. Science 212: 1015–1017.

Mintz, B., and Illmensee, K.
   1975   Normal genetically mosaic mice produced from malignant teratocarcinoma cells. Proc. Natl. Acad. Sci. 72: 3585–3589.

Papaioannou, V. E., McBurney, M. W. Gardner, R. L., and Evans, M. J.
   1975   Fate of teratocarcinoma cells injected into early mouse embryos. Nature 258: 70–73.

Probber, J., and Ehrman, L.
   1978   Pertinent Genetics for Understanding Gender. In Genes & Gender, Tobach, E., & Rosoff, B. (ed.) N.Y., Gordian Press, 1978, pp. 13–30.

WHAT'S NEW IN ENDOCRINOLOGY? TARGET: SEX HORMONES

Susan Gordon, M.D.

Women, simply as women and not necessarily as scientists, have always had an interest in their bodies. They have wondered about the obvious anatomical differences between men and women, about menstrual periods, conception, pregnancy and childbirth. Interest in other biological systems has been high, also, but probably not as intriguing or intense as in the reproductive system and its function.

Paralleling recent research in the field of genetics has been the ongoing study of hormones, particularly sex hormones. In the continuing quest for understanding the differences between men and women, investigation has long been focused on the study of these hormones as well as of the sex chromosomes . . . this, in the hope of explaining sexual dimorphism scientifically.

This paper attempts to update and clarify some of the recent research in the area of reproductive endocrinology. It endeavors to pick up where Anne Briscoe's article in Genes and Gender leaves off. (Briscoe, 1978) Since it is impossible to cover all the research, I have tried to choose those areas that might be of more interest and/or those that may have more information available about them.

Very little of the work has been done on human beings and it behooves us to remember, as Ruth Bleier points out in her beautiful paper in Genes and Gender II, that the vast amount of endocrine research is in the area of sex hormones and by and large most of it is done in rats or other lower animals. The research is done mostly by men and men in industrialized, western capitalist nations with patriarchal societies. (Bleier, 1979) Although we may be able to extrapolate some of the data from rats to human beings, particularly in relationship to substance and mechanism, the overall expression of the effect of the hormone itself may be quite different in people given the fundamental differences between the human and rat brain.

First we do know that in the development of the embryo, without some testicular determining substance, the embryonic gonad (sex gland) will develop into an ovary. Also, the hypothalamus (a part of the brain) will retain the female cyclic pattern (monthly ovulation) unless it is changed by some substance, as yet not definitely defined in the male.

The Y chromosome (the male chromosome) at about the 6th week of intrauterine life directs the formation of the testes. This appears to be the predominant mode of sexual differentiation in the human and other primates and occurs because of the presence of a gene product known as the H-Y antigen. Present evidence points to regulators of the production of this antigen by genes on the Y chromosomes but its expression and function are subject to other influences, one of which may be an X-linked gene. (Gordon and Ruddle, 1981)

Also, there are some reported examples of testicular development and positive H-Y antigen determination when chromosomal evidence of Y is missing, in a mouse, at least! (Haseltine and Ohno, 1981)

In general, then, H-Y antigen guides the differentiation of the gonadal ridge to the fetal testis. Otherwise, the gonadal ridge would become an ovary. There is not yet conclusive evidence that the H-Y antigen is *the* testis-determining substance but it is a widely held working hypothesis that it is so. (Gordon and Ruddle, 1981)

The fetal testis then begins to secrete 2 substances: the classic male hormone, testosterone and a Mullerian inhibiting substance. These two substances masculinize the fetus anatomically. That is, they are largely responsible for the formation of the penis, scrotum, prostate gland, epididymis, vas deferens and seminal vesicles. In the absence of these hormones the fetus develops female structures, i.e., uterus, Fallopian tubes, vagina, clitoris and the labia.

Furthermore, fetal testosterone seems to be a major factor in masculinizing and/or defeminizing the hypothalamus so that later, as puberty approaches, male maturation can occur. That is male secondary sex characteristics develop and reproductive capacity (the production of sperm) is established. In the female it is not clear whether sexual differentiation at maturity takes place simply because of the relative absence of testosterone or whether there are hormonal factors in the female that protect her from the differentiating effects of androgens (Maclusky and Naftolin, 1981). In any case, whatever potentiates the hypothalamus in male or female the neuroendocrine axis is basically the same. It consists of the following pathway:

$$\text{hypothalamus} \xrightarrow{\text{GnRH*}} \text{pituitary} \xrightarrow{\text{GTH(LH\&FSH)**}} \text{gonads} \xrightarrow{\text{androgens \& estrogens}}$$

Once this stimulant-feedback mechanism is set into motion, secondary sex characteristics appear, ovulation or sperm production begins and the reproductive capacity of the individual is in place.

Investigators have been quite intrigued with the whole area of masculinization and/or defeminization of the brain in mammals including human beings. Thus a great deal of recent research deals with this whole aspect of how masculine central nervous system differentiation is achieved. Here a few definitions are in order. MacLusky and Naftolin (MacLusky and Naftolin, 1981) define defeminization as "suppression of the behavioral and neuroendocrine patterns characteristic of the female" and masculinization as "enhancement of the patterns characteristics of the male." (Note the inclusion of *behavioral* as well as neuroendocrine characteristics.) In human beings, and certain monkeys it (masculine brain differentiation) is accomplished through the androgen (testosterone and dehydrotestosterone (DHT)) pathway acting directly on receptor brain cells including cells of the hypothalamus. This is the main pathway. In other mammals such as the rat and hamster, it has been shown that testosterone is changed into an estrogen (the classic female hormone) specifically estradiol, in the brain cells through a process called aromatization. According to some investigators this process also takes place in the human. Present in both male and female brain cells are receptors for both androgens and estrogens (McEwen, 1981) which helps to sup-

* GnRH—Gonadotropin Releasing Hormone
** GTH—Gonadotropic Hormones
** LH—Luteinizing Hormone
** FSH—Follicle Stimulating Hormone

port the above hypothesis but adds another dimension to the relationship of "male" and "female" hormones.

Why is not the male fetus feminized by his mother's "female" hormones? This is another intriguing area of investigation. In some animals a certain substance called alpha feto-protein seems to be the protective substance, but in humans, the mechanism has not yet been completely worked out although it is believed that alpha feto-protein plays a role in humans, too.

Still another area of recent research which has inspired tremendous interest is an elaboration of the endocrine mechanism which triggers ovulation. This mechanism is well on its way to being understood. We know now that the hormone called gonadotropin releasing hormone (GnRH) also known as luteinizing hormone releasing hormone (LHRH) is elaborated by cells located in certain areas of the hypothalamus of the brain. GnRH creates a surge of luteinizing hormone (LH) from the anterior pituitary gland about the midpoint of the menstrual cycle which causes the egg to be released from the ovary. What we don't know, however, is what causes the surge. Is it the result of increased GnRH secretion or an increased pituitary response or sensitivity to GnRH? Present research is focused on these two questions. Another interesting aspect of the LH surge is that it can be detected by finding certain chemicals in the urine. The *in vitro* fertilization of a human egg (the so called "test tube baby") technique, is based on quantitating these chemicals. The technology for a simple home test for detecting the LH surge is feasible and would be a safe and accurate birth control method.

GnRH also causes ordinary amounts of LH and follicle stimulating hormone (FSH) to be released from the pituitary gland. These in turn stimulate the ovary to secrete hormones; estrogens, progestins (the female hormones) and androgens (male hormone). Estrogens and progestins in addition to anatomically and physiologically feminizing the individual, feed back to the brain and pituitary gland and keep the cycle going. This is known as the neuroendocrine axis or the hypothalamic-pituitary-ovarian axis. A similar mechanism occurs in the male, already partially alluded to previously in this paper. That is that GnRH causes the pituitary gland to secrete LH and FSH which control the function of the testes specifically to secrete male hormones, (androgens), and to manufacture sperm. The main difference between males and females is the absence of a surge of LH in the male. This is the result of masculinization and/or defeminization of the fetal brain.

What sets off this neuroendocrine mechanism in the first place? What role does the pineal gland, also located in the brain, play? Generally, its hormone, melatonin, acts to decrease the amount of LH and FSH (gonadotropins). Since this secretion, melatonin, an anti-gonadotropin hormone, is synthesized at a rate inversely dependent on environmental lighting perhaps the degree of exposure to light is the key factor in starting off the mechanism that leads to ovulation. One statistic that bears out this hypothesis is the following: In northern Finland where there are very long nights in winter and very long days in summer, the incidence of conception is lowest in January and highest during the spring and summer. It is also known that many menstrual cycle irregularities have been stabilized following deliberate night time exposure to light during the mid cycle. Other evidence that the pineal gland plays a major role in reproduction is borne out by the fact that tumors of the gland may produce either sexual precocity or delayed maturation of the sex organs (gonads). This mysterious gland probably also plays a role as one of the major reproductive controls in animals that breed seasonably. (Martin, 1976)

Another hormone about which we know very little is prolactin. This hormone has recently come under greater study and scrutiny. Its only previously established role is the initiation and maintenance of lactation. New roles are being defined such as a role in stimulation of the ovary along with LH and FSH to secrete steroid hormones (estrogens and progestins), in inhibiting implantation of the fertilized egg, in influencing carbohydrate, water and electrolyte metabolism and at least some role in the onset of puberty. In the male it is important in the growth and development of the accessory reproductive structures such as the prostate and often manifests itself during puberty through the temporary production of milk in the male adolescent. (Incidently, "dimorphism of breast development has not been documented in the human embryo and the breasts of boys and girls are identical prior to the onset of puberty," (Wilson, George and Griffin, 1981) Most recently, prolactin has been implicated as a mediator in a number of estrogen effects on neuro-transmitter turnover in the brain. (McEwen, 1981)

A third sex hormone that we know very little about is inhibin. This hormone seems to be involved in inhibiting FSH production in the male, thus exerting some effect on sperm production. It is thought to be the FSH inhibitor in ovarian follicular fluid, too, which acts to enhance the feedback effect of estrogens in the female. At one time it was hoped that inhibin might be an effective and safe male contraceptive. (Main and Davies, 1979) Until total isolation of this hormone and better understanding of its functions take place, that research cannot begin.

Other organs have a role in reproductive function and an effect on sex hormones: the liver, the adrenal and thymus glands and fat (adipose) tissue. The liver breaks down (detoxifies) sex hormones and synthesizes some. The thymus role is very weak and still quite unclear. The adrenal produces both androgens and weak estrogens. Adipose tissue converts the weak adrenal estrogen into stronger estrogen. (Martin, 1976, b, c, d.)

Thus we can see the complex nature of the production and utilization of our sex hormones. Nature has provided a safety valve by having many organs and systems involved in the entire process. It is also evident that both males and females produce both "male" and "female" hormones and both have receptors for both. Even males produce prolactin for instance, and inhibin is found in ovarian follicular fluid. These are just two examples. Others have been mentioned previously. Hence we are all truly biologically both male and female. Sexual dimorphism depends to some extent on genetic processes, quantity of hormone and to a great deal on cell and organ system receptivity, environment and total organism milieu. Even the so-called "almighty" gene is subject to varying expressivity in different individuals subject again to internal (i.e., hormonal and other processes) and external environmental conditions. Neither the actions of the hormones nor of the gene would be expressed, if there were no environment.

The next question to be answered is: Do sex hormones determine the behavior of human beings? Much research has been done to show that sex hormones do indeed determine behavior (sexual behavior in rats, at any rate). The influence of sex hormones on behavior of lower animals (particularly rats) is measured by reproductive behavior, i.e., mounting behavior in males and lordotic positioning in females. In humans to prove the effect of pre-natal sex hormone on sex-dimorphism, six areas of behavior have been studied in children.

(1) Physical Expenditure—meaning intense, active outdoor play.

(2) Social aggression—physical and verbal fighting.

(3) Play rehearsal of parenting behavior—measured as playing mother and father, participating in infant care and playing with dolls.

(4) Effects on peer preference.

(5) Gender role labeling—"tomboy" or "sissy."

(6) Grooming behavior—adornment as indicated by clothes preference, jewelry, make up, hairdo, etc. (Ehrhardt and Meyer-Bahlburg, 1981)

In essense, these are not comparable to the behavior described in animal studies and in measuring them there is no clear cut evidence of the action of the hormones. In fact, the choosing of these areas of behavior to measure, is in itself, sexist, and hardly in accordance with modern societal practices.

The post-natal sex hormone effect on human behavior is even less conclusive than the pre-natal effects.

The areas of behavior that have been studied are:

(1) testosterone and aggression in men

(2) mood and the menstrual cycle in women

(3) pubertal sex role reversal in pseudohermaphrodites

The conclusions of the authors of a recent paper on the subject were as follows "the foregoing examples of postnatal gonadal steroid effects on sexually dimorphic behavior in humans highlight the complexity of the interaction between hormonal and psychosocial factors" ". . . present day knowledge indicates that in each area (of behavior researched) there is a complex interaction between hormonal influence on brain function and psychosocial and environmental forces that results in the expression of sexually dimorphic behavior. Thus, the concept of a "nature versus nurture" dichotomy is anachronistic; the goal of future research should be to delineate the relative contribution of both sets of influence on each specific behavior." (Rubin, Reinisch and Haskett, 1981)

What other points should be made?

(1) All mammals share the physiology, biology, biochemistry and anatomy of sex to a great extent, but behavior differs from one species to another, from one individual to another. Given the difference between a rat brain and a human brain, for instance, it is obvious that with the human ability to use symbols, rationalize, create, abstract and synthesize there is a vast difference in the behavior of the two species so that sexual behavior in the human is greatly influenced by the socialization and experience of the individual as well as her/his ability to think.

(2) One can question why so much research is being done in the area of sexual dimorphism with its particular emphasis on gene "programming" and male and female behavior as determined by hormones and genes, and the extrapolation of this information on rats, guinea pigs and monkeys to men and women. Is this not sexism and an attempt to continue the concept of genetic determinism and thereby, racism too?

(3) Lastly, this kind of research deters us from focusing on other more important problems such as those having to do with health and well-being, not only of women but of all humankind. For example, why not research in the area of mood changes in men as well as women? Let's find out more about the physiology of menopause in women. What about menopause in men? Wouldn't it be interesting and helpful to know what happens to hormones in women athletes who no longer menstruate or those who suffer from anorexia nervosa? Shouldn't we be looking into how men, women and children cope with present day anti-human onslaughts, be they toxins in our environment or emotional economic and political stresses, (to name a few). All of these projects which lead to the betterment of human beings and improve the quality of life for all is where we should be putting our research efforts, not into those areas which continue to beat the drum of racism and sexism.

REFERENCES

Adler, H., ed.
1981 *Neuroendocrinology of Reproduction*, New York and London: Plenum Press.

Bleier, R.
1979 Sexual and political bias in science; an examination of animal studies and their generalizations to human behavior and evolution in *Genes and Gender II*, Hubbard, R., Lowe, M., (eds.), New York Gordian Press, 49–71.

Briscoe, A.
1978 Hormones and gender in Genes and Gender I, Tobach, E., and Rosoff, B., eds., New York Gordian Press, 31–51.

Ehrhardt, A., and Meyer-Bahlbur, H.
1981 Effects of prenatal sex hormones in gender-related behavior. *Science, 211;* 1312–1317.

Federation Proceedings
1980 *Symposium on Puberty, 39,* 7.

Gordon, J., and Ruddle, F.,
1981 Mammalian gonadal determination and gametogenesis. *Science, 211,* 1265–1271.

Hasseltine, F., and Ohno, S.
1981 Mechanism of gonadal differentiation. *Science, 211,* 1272–1277.

Ingbar, S., ed.,
1979 Prolactin in *Contemporary Endocrinology*, New York Plenum Press, *1,* 38–46.

Main, S. and Davies, R.
1979 Inhibin—and endocrine enigma. *Trends in Biochemical Science.*

Martin, C.
1976 The Pineal and thymus gland, *Textbook of Endocrine Physiology*, Baltimore Williams and Wilkins a. 408–412, b. 253, c. 74, d. 253, 436–457.

Maclusky, N., and Naftolin, F.
1981 Sexual differentiation of the central nervous system. *Science, 211,* 1294–1303.

Mc Ewen, B.,
1981 Neural gonadal steroid actions. *Science, 211,* 1303–1311.

Nolin, J.
1978 Target cell prolactin in *Structure and Function of Gonadotropins*, McKerns, K., ed., New York Plenum Press, 151–182.

Rubin, R., Reinisch, J., and Haskett, R.
   1981   Postnatal gonadal steriods effects on behavior. *Science, 211*, 1318–1323.

Shapiro, B., Levine, D., and Adler, N.
   1980   The Testicular feminized rat: a naturally occurring model of androgen independent brain masculinization. *Science 209*, 418–420.

Turgeon, J.
   1978   Neural Control of Ovulation, *Tutorial Lecture, Fall Meeting of the American Physiological Society.*

Wilson, J., George, F., and Griffin, J.
   1981   The Hormonal control of sexual development. *Science, 211*, 1278–1284.

SEX AND TEMPERAMENT REVISITED

Jagna Wojcicka Sharff, Ph.D.

Over a half a century ago, a tiny young woman sailed away alone on the S.S. Sonoma, bound for the Pacific Islands. She had managed to convince her family, friends and advisor that in spite of being fragile and a female, she was capable of living and working among "primitive people." Later, in her *Letters from the Field* (1977) and in her autobiography (1972) she was to describe the difficulties she had to overcome in order to obtain funds and the sceptical approval for her pioneering work in a field which was then dominated by men and their theoretical concerns. By then, Margaret Mead had become the most widely known ethnographer.

Her most important book, published during the early years of her research was *Sex and Temperament* (1935). This work challenged stereotypic notions of sex linked behavior. Comparing the sex roles in three societies which she had studied, the Arapesh, the Mundugumor and the Tchambuli, she concluded:

> Neither the Arapesh nor the Mundugumor profit by a contrast between the sexes; the Arapesh ideal is the mild responsive man married to the mild responsive woman; the Mundugumor ideal is the violent aggressive man married to the violent aggressive woman. In the third tribe, the Tchambuli, we found a genuine reversal of the sex attitudes of our own culture, with the woman the dominant, impersonal managing partner, the man the less responsible and the emotionally dependent person. (1963:279)

She concluded that: ". . . we no longer have any basis for regarding such aspects of behavior as sex linked." (Ibid.:280).

Mead made that statement nearly fifty years ago. Since then important information has been accumulated confirming her insight. Yet, even today the issue of gender and behavior is being confounded. Recurrently, in response to the ideological needs of the time, male writers assert or imply a biological basis for male supremacy.

In this paper I will first review some of this recent literature and show that it lacks both a historical dimension and an ecological perspective. Next I will present some evidence from mine and others' work to demonstrate the fact that sexual roles are flexible and evolve in adaptation to basic socio-economic conditions. I will conclude by suggesting that the ideological insistance of biological determinism of women's domestic roles will have grave consequences for the status of women's health.

The revolutionary decade of the thirties during which *Sex and Temperament* was published was followed by the Second World War and the intellectual freeze of the fifties. As women returned to the "hearth" to tend babies and perform domestic service, social scientists gave strong support in their writings to the supposed passive and responsive nature of women. Among them unfortunately, was Margaret Mead. In her *Male and Female* published in 1949, she blurred the initial import of her former work by describing the prevalent *status quo* in sex roles as natural and desirable. She writes, for example, "Success for a woman means success in finding and keeping a husband. This is much more true than it was a generation ago when . . . some women found their new freedom outside the home so intoxicating that they could *abandon* themselves to their work" (1967:324; Italics added). In sentiments such as this one pronounced throughout this book, Margaret Mead not only deprecated her own worth, as a consummate, non-traditional female, a living example of the meaninglessness of assigned sex roles, but she also helped to whet the appetites of male social scientists who were ever ready to reassert the vision of a perpetually subordinate domesticated female.

Every twenty years or so, a new crop of publications appears purporting to show definitively that gender roles of our society are biologically based. The decade from the mid-sixties to mid-seventies was very fertile in that respect. We are all familiar with highly publicized books such as the works of Desmond Morris (1968) and Konrad Lorenz (1966), both scientists who do not specialize in the study of human behavior, popular authors such as Ardrey (1961, 1966) and even anthropologists, specifically Tiger (1970) and Fox (1971). Edward Wilson, (1975) a noted authority on insects added a new dimension to the discussion by explicitly proposing the genetic determination of human behavior.

Yet another anthropologist, Napoleon Chagnon, has recently joined the ranks of sexual determinists. His work merits some discussion because it appears to be based on solid ethnographic data and has been widely disseminated through popular publications and the media. In his first ethnographic work, *Yanomamö: The Fierce People,* Chagnon (1968) describes a hunting horticulturalist group of Indians who reside in shifting villages in the Amazonian jungle basin. He writes "Yanomamö society is decidedly masculine" (1968:81) and provides a vivid description of the brutality practiced by men upon women which includes slapping, beating, wounding, maiming and on occasion, killing. He asserts that the dynamics of this male supremacist complex involve female infanticide, hence a scarcity of women that is further intensified by the polygamous monopoly of women by older men.

In addition male control over female labor and hostility among male kinsmen and between villages leads to a pattern of incessant warfare, raiding for women and theft of garden produce. Men must continuously clear new garden sites and forge new alliances as villages fission and move in response to internal hostility and warfare. Inter-village alliances are easily, and often treacherously, broken.

Chagnon (1968:123) writes, "New wars usually develop when charges of sorcery are leveled against the members of a different group." But once the group is on the warpath the raiders hope to acquire women. Furthermore "Most wars are merely a prolongation of earlier hostilities stimulated by revenge motives . . . the first causes of hostilities are usually sorcery, murders or club fights over women in which someone is badly injured or killed."

In an effort to explain the seeming irrationality of this male supremacist, warring complex, anthropologist Marvin Harris (1975) has proposed that an explosive rise in population among the Yanomamö during the past hundred years, resulting from an increased dependence on the newly introduced plantains and bananas caused a permanent depletion of the wild animals in the area. He writes "There is evidence that the Yanomamö have 'eaten' a very important resource as a result of population expansion" (276) and that "they are now struggling with each other to gain access to the remaining hunting areas on the margins of their former territory." (279).

Chagnon (1974:127) subsequently refuted Harris' ecological explanation by writing of "enormous tracts of land abounding with game" within Yanomamö territory. However this contradicted earlier statements which said in part: "Game animals are not abundant and an area is rapidly hunted out so that a group must keep constantly on the move . . . I have gone on five day hunting trips with the Yanomamö in areas that had not been hunted for decades . . . and we did not collect enough food even to feed ourselves." (But) "on other trips we often managed to collect enough game to feed the entire village." (1968:33). It is difficult to evaluate Chagnon's assertions because he not only presents us with contradictory statements but also does not provide any quantified or quantifiable information which could settle the point.

Similarly, in describing woman's and man's work Chagnon is contradictory and impressionistic. Although we are told that a woman's duties as a wife "require her to assume difficult and laborious tasks too menial to be executed by the men" (Ibid.:81) a close reading of the text reveals a different picture. For the most part these laborious menial duties consist of fetching water, foraging for firewood, playing with children and doing *some* of the cooking. To be fair to Chagnon, he obviously does not give us the full inventory of women's work; it has to be inferred between the lines. He only stresses what he considers menial.

Meanwhile, the Yanomamö men not only have to go on exhausting hunting and warring trips* but they also prepare food, cook for large numbers of people, manufacture tools, weapons and drugs and also do most of the farming. Consider the following passage:

---

* Boys as young as twelve years of age may be pressed into war parties. (1968:132) And many young men "chicken out" from hunts and wars, return back to the village sheep-faced with imaginary illnesses after several days (passim).

> All day long Kaobawa's younger brothers who had returned from the hunt the day before labored at cooking the enormous quantity of ripe plantains, pouring each boiling containerful into the trough as it was prepared . . . They worked at this task from early morning until late afternoon, in addition to boiling a nearly equal quantity of green plantains. (Ibid.:108).

And on the days when they were not hunting, warring or cooking the easygoing male Yanomamö daily life looks as follows:

> Within an hour after it is light the men are in their gardens clearing brush, felling large trees, transplanting plantain cuttings, burning off dead timber or planting new crops of cotton, maize, sweet potatoes, yucca, taro or the like, depending on the season.

Lest the reader begins to wonder about which sex works harder, Chagnon follows this passage with: "Little girls learn very quickly that this is a man's world for they soon must assume much of the responsibility for tending their younger siblings, hauling water and firewood and in generally helping their busy mothers." (Ibid.) Is this a way for a girl to learn that this is a man's world? Or is it a way for us to learn that Chagnon did not describe what women really do and only labelled what he did notice as "difficult, laborious and menial"?

Further close reading reveals that neither young women nor young men have much authority in this society. The stature of both sexes increases with age. Furthermore Chagnon himself notes that women as a group exercise considerable influence by concerted goading of men into wars (Ibid.:84) and into extended hunting for scarce meat. A returning hunter may even refuse to eat a portion of his catch so that the women and children may have more. (Ibid.:91–92).

Thus Chagnon himself contradicts his picture of the fierce Yanomamö on which some recent grandiose theories of male dominance are being built. Questions about Yanomamö gender are not purely academic. In his most recent public pronouncements Chagnon has alligned himself philosophically with sociobiology. An article in *Time* magazine entitled "Beastly or Manly?" states "Implied in Chagnon's findings so far is a notion startling to traditional anthropology: the rather horrifying Yanomamö culture makes sense in terms of animal behavior." After comparing the Yanomamö to "baboon troops" the article quotes Chagnon as saying: "In primates and all mammals internal social organization results from the breeding system and there's no reason to believe it's not true of humans. It's possible that war and marriage make sense in zoological terms and Darwinian theory is applicable to human behavior." (quoted in Davis, 1976). And finally, in a National Geographic article Chagnon wrote:

> I have gradually come to realize that a chronic shortage of women *determines* much of these Indians' social structure. One theory in anthropology is that warfare among primitive peoples can be usually traced to conflicts over land or water or some other strategic resource. Another view holds that blood relatives do not war against each other. The Yanomamö refute both theses by their actions (quoted in Davis 1976).

In the above passage, Chagnon in hot pursuit of sexual determinism, fails to spell out the crucial importance of land and does not bother to mention the other strategic resources important to analyzing the problems facing the Yanomamö.

In his original ethnography, Chagnon asserted that at the time of his research "many villages have not yet been contacted by outsiders" (1968:1). Yet there is evidence that some contact did exist especially in the area in which he worked. Chagnon himself reveals this fact, often in footnotes. In the first few pages of his book we discover that Chagnon was accompanied on his first visit to the Yanomamö by a Mr. James Barker, a Protestant missionary who had been working with the Yanomamö for the past fifteen years (1968:4). Next we discover that there are a number of missionaries working in other Yanomamö villages in the area (1968: footnotes pp. 9, 34, 71, 78, 122). We also learn a very important fact from a footnote, i.e., that among the Yanomamö "two percent of all adult deaths are due to snake bite; 54 percent are due to malaria and other epidemic diseases and 24 percent of adult males die in warfare" (1968:20). It does become apparent, therefore that many of the Yanomamö groups had had contact with "outsiders" through missionaries, through trade and indirectly through exposure to infectious diseases.

In fact, historical sources suggest that the Yanomamö had been driven into their present shrinking habitat by forces of colonial expansion. According to Davis:

> For nearly a century, it appears as if the Yanomamö were forced to retreat defensively into their present territory between the Orionco and Marauia Rivers. To the south, they were attacked by Brazilian rubber collectors and settlers. To the north, they fought off the expanding cattle frontier in Venezuela, and the more acculturated and rifle-bearing Makiritare tribe (1976:12).

We can now see the Yanomamö as retreating defensively into a smaller land base at the same time as they expand numerically as a result of their new reliance on staple cultivation. In an environment depleted of faunal resources they are in constant struggle for new garden sites and hunting territories. One can go even further and suggest that Yanomamö society is in a state of crisis resulting from colonial contacts. According to Lizot the introduction of metal tools and shotguns created a disequilibrium among the contacted and the more isolated Yanomamö groups. He writes: "The entire map of economic and matrimonial circuits along with political alliances was transformed and flagrant imbalances developed" (Quoted in Davis, 1976). It is therefore more than probable that a fierce territorial struggle contributed to the development of the sexist warrior complex which Chagnon described.

That Yanomamö society is not universally fierce, and may not have always been, can be deduced from an account by another ethnographer. Smole, who worked among a relatively peaceful and isolated highland Yanomamö group reports that male/female relationships among them are much more egalitarian than those reported by Chagnon (c.f. Leacock, 1978). Similarly, a recent ethno-historical account of an Amazonian society in transition suggests that male dominance described for several of the Amazonian groups may have been the result of contact and forced accommodation to Western imperialism. The authors write:

> . . . while the classless nature of Amazon horticultural societies has been frequently reported, such groups are not generally described as harmonious and peace loving. In fact studies of South American groups that have been more extensively exposed to Western imperialism such as the Yanomamö, the Mundurucu and the Jivaro indicate quite the opposite (Buenaventura-Posso and Brown, 1980:115).

The authors of the study describe the historical change among the Bari of Colombia from egalitarianism toward male dominance. On a brief visit to this society they were impressed by the "egalitarian and gentle" relations between the sexes but note that the situation is changing:

> The current impositions of the market economy, combined with the growing infringements of Colombian and Venezuelan national and international missionaries are beginning to affect relations between the sexes. These relations unfortunately are as vulnerable to the forces of ethnocide as other cultural components . . ." (Ibid.:130).

Similar ethnohistorical and ethnographic research in such diverse areas as South and North America, Africa and Australia (Etienne and Leacock, 1980) reveals the following cross-cultural insights. In egalitarian societies based on hunting and gathering men and women are autonomous with respect to work performance and even ritual. Their economic and social contributions to the welfare of the group are equally valued and respected. There is no differentiation of "public" or "private" domains often associated in our society with male and female work respectively. As Leacock writes:

> . . . in egalitarian society a "private" familial, female domain is not defined and made secondary to

a "public," political male domain. Instead, authority is dispersed and decisions are by and large made by those who will be carrying them out. All manner of social arts are used by both women and men to influence people, resolve problems and hold groups together. These range from endless talk and discussion through myth making, song, dance and ritual to merciless teasing, disapproval and threat of social isolation (Etienne and Leacock, 1980).

This research goes beyond mere documentation of the male and female complementarity in egalitarian societies. It suggests that even in horticultural ranking societies of pre-European contact time, women retained a great deal of autonomy and power through control of their own productive and reproductive activities. The most important point that emerges from this work is that the relations of male dominance now encountered in so many societies were not necessarily indigenous developments. The agents of colonialism; the missionaries, the military, the administrators, and currently the multinational business and agrobusiness, actively encouraged and fostered the development of hierarchical relationships in order to consolidate their control over previously more egalitarian societies. Leacock's (1981) work in particular illustrates the importance of a diachronic or historical approach to the understanding of sex roles, the way they were structured, how they were transformed and also how they can be re-structured in the future.

In my own research with migrant Puerto Ricans I utilized a diachronic, cultural materialist approach to define the economic, political and historical determinants of sex roles. In a four year study conducted on the Lower East Side of Manhattan (Wojcicka Sharff, 1981) I discovered that in spite of the assertions by both men and women of normative male superiority, the observable or etic reality was quite different. In the historical context of colonization and of internal migrations from mountain subsistence farming, to sugar crop wage labor, and to industrial wage work on the island and on the mainland, men and women experienced a series of role and status changes, modifications and reversals. This occurred within the life time of the men and women I worked with, often within a period of a few years, or a few months. I found that for poor urban Puerto Rican New Yorkers, role and status are determined by the importance of the person's contribution to the survival of the household, and thus ultimately, to the person's position in the economy. Since women can gain a livelihood more easily than men they emerge as the heads of vigorous, female based households. Furthermore, in such situations, women tend to "invest" in their children by preparing them, through differential socialization, for various roles that will assure the viability of the households.

During my research I have observed at least four such roles for which the children are reared. These include what I have termed the "street representative," the "wage earner," the "child reproducer" and the "scholar/advocate." The children are encouraged to develop appropriate qualifications along the male-female continuum to prepare them for each of these adult roles.

The street representative is raised to be a "macho man," because his function will be to perform the dangerous work of defending and avenging family members, and pursuing activities which are sometimes illegal, and often risky. The wage earner, usually a male, develops passive qualities, to prepare him for wage labor, while the child reproducer is encouraged to develop traditionally feminine qualities of gentleness and succorance. The scholar/advocate, on the other hand, who is usually also a female, begins early in her life to train for dominance and assertiveness. The entire household contributes to this girl's eventual upward mobility.

My research suggests that sex roles are malleable and that they respond, through enculturation to the opportunities available within the larger economy. In a situation of underemployment and inadequate social support for those unable to work or those unable to find work, the response on the household level is to mobilize all members into income generating activities. Mothers expand their support networks in the present and provide for their own "social security" in old age by investing in children. Some children are encouraged to develop qualities usually associated with the opposite sex to prepare them for the opportunities existing in the labor market.

In conclusion I would like to suggest that re-mystification of sex roles helps to cover up the exploitiation of women's labor.

This has serious consequences for women's health, particularily in third world countries. A recent United Nations study on the status of women noted that the economies of most countries have come to rely on large numbers of women in the workforce, without giving adequate consideration to the health effects on women and their families. It states in part, "Increasingly women in developing countries are being used as a cheap source of labor" and that "the developing countries are likely to reproduce the health crisis of the industrial revolution in 19th century England, with women as the most vulnerable victims." (United Nations 1980:4). A major reason for this pessimistic prognosis, the report states, is that women engage in "double day" loads, that is, they must work in wage labor and perform domestic tasks. One of the major obstacles to alleviating the work load, the report notes are the traditional stereotyped images of men's and women's proper roles. As a consequence of this women are held responsible for domestic duties, regardless of their work outside the home.

As anthropologists, we can document that the attributes of "passivity" or "aggressiveness," or "public" or "private" spheres of competence vary cross-culturally and across time, that indeed, in Margaret Mead's words "we no longer have *any* basis for regarding such aspects of behavior as sexlinked" (1968:280). I believe that as we learn more about our common egalitarian heritage, and as we begin to discover "human nature" in its manifold plasticity we can begin to lay a basis for a future in which women and men can cooperate on the basis of autonomy, complementarity and mutual respect.

REFERENCES

Ardrey, Robert
    1966   *The Territorial Imperative.* N.Y. Atheneum.

Buenaventura-Paso, E. and S. E. Brown
    1980   Forced Transition from Egaliterianism to Male Dominance: The Bari of Colombia. In Etienne M. and Leacock, E., eds. *Women and Colonization,* N.Y. Preager.

Chagnon, Napolean
    1968   *Yanomamo: The Fierce People.* N.Y. Holt, Rinehart and Winston.
    1974   *Studying the Yanomamo.* N.Y. Holt, Rinehart and Winston.
    1976   Yanomamo: The True People. National Geographic, 150, 2.

Davis, Sheldon
    1976   The Geological Imperative: Anthropology and Development in the Amazonian Basin of South America. Boston, MA. *Anthropology Resource Center.*

Etienne, Mona and Leacock, Eleanor
    1980   *Women and Colonization.* N.Y. Preager.

Harris, Marvin
    1975   *Culture, People and Nature.* N.Y. Thomas Y. Crowell.

Leacock, Eleanor
    1978   Society and Gender, in *Genes and Gender,* Tobach, E. and Rosoff, B., eds. N.Y. Gordian Press.
    1981   *Myths of Male Dominance: Collected Articles on Women Cross Culturally.* N.Y. Monthly Review Press.

Lorenz, Konrad
    1966   *On Aggression.* N.Y. Harcourt Brace.

Mead, Margaret
    (1935) *Sex and Temperament in Three Primitive Societies.* N.Y. Morrow
    1963   Quill Paperbacks.
    (1949) *Male and Female. A Study of the Sexes in a Changing World.* N.Y.
    1967   Morrow Quill Paperbacks.
    1972   *Blackberry Winter: My Earlier Years.* N.Y. Simon & Shuster.
    1977   *Letters from the Field 1925–1975.* N.Y. Harper & Row.

Morris, Desmond
    1968   *The Naked Ape.* N.Y. McGraw-Hill.

Smole, William
    1976   *The Yanomamo Indians: A Cultural Geography.* Austin, Texas, Austin University of Texas Press.

Time
    Manly or Beastly? May 10, 1976.

United Nations
  1980   Worsening Situation of Women Will Be Main Issue Confronting Commission on the Status of Women. *D.P.I./D.E.S.I.* Note IWD/ 22 13 February.

Wilson, E. O.
  1975   *Sociobiology: The New Synthesis.* Cambridge, The Belknap Press of Harvard University Press.

Wojcicka-Sharff, Jagna
  1979   *Patterns of Authority in Two Urban Puerto Rican Households.* N.Y. Columbia University doctoral dissertation.
  1981   Free Enterprise and the Ghetto Family. *Psychology Today,* March.

The previous papers provided scientific documentation for views which are contrary to popularly held notions. Because a belief is commonly held, many come to regard it as 'true' and never question the basis on which it is formed.

The following three papers attest to the effect of such ideologies on women's health in the workplace. Harris discusses a particular example of occupational disability, the so-called assembly line hysteria, and raises questions about what kind of problem it is. He further demonstrates that the manner in which information about the concept is disseminated tends to reinforce assumptions about women's responses. Bellin and Rubenstein write about the harmful effects of chemical substances on human physiology. They explain why definite proof of adverse effects is often hard to obtain and criticize management's tendency to get rid of the endangered workers—specifically the women among them—instead of the hazardous conditions. The last article is taken from a curriculum guide for workers put together by the Reproductive Hazards Committee of the New York Committee for Occupational Health and Safety (NYCOSH). The authors graciously gave us permission to use it. The article deals with the process of participatory education and with ways of putting the information obtained to its most effective use.

THE MYTH OF ASSEMBLY-LINE HYSTERIA

Ben Harris, Ph.D.

This paper addresses the problem of "assembly-line hysteria." Over the last two decades, press reports and journal articles on this problem have dramatically increased in number (see Sirois, 1975, Table 1). According to these reports, assembly-line hysteria is an infrequent but troublesome disorder affecting groups of women workers. As the reported frequency of this disorder has increased, it has become a concern of labor, management, and of agencies such as the National Institute of Occupational Safety and Health (Note 1). Because women workers are reported to be the primary victims of assembly-line hysteria, I suggest that this problem also deserves the attention of anyone concerned about women's economic and biological rights—or about the ideology of those who attack these rights.

In this essay I describe what people have in mind when they refer to the concept of "assembly-line hysteria," how this concept has been analyzed by psychologists and sociologists, and how it has been used by the press. Then I discuss some serious problems with current applications of this concept in industry. Although I have titled my essay "The *Myth* of Assembly-line Hysteria," I leave it to the reader to judge whether I'm describing a real creature, a mythological one, or perhaps simply a mysterious one (Colligan & Stockton, 1978). I acknowledge the existence of the "problem" of assembly-line hysteria; at the same time, I raise the question of what *type* of problem it is (e.g., medical, disciplinary, psychological, imaginary), and *by whom* it is considered a problem (individual workers, psychologists, management).

Definition

The majority of researchers in the United States currently use the term "mass psychogenic illness" (MPI) to refer to what others call "mass hysteria," "mental epidemics," and "multiple occurrences of unexplained symptoms." While some theorists have complained about theoretical inaccuracies related to the term MPI (see Markush, 1982; McGrath, 1982). I will use it here as a synonym for all of the above-listed terms. In doing so, I will broadly define it as: the relatively sudden occurrence in more than one person of the signs of illness (e.g., headache, nausea, vomiting, fainting) without an identified physical cause (see also Colligan & Murphy, 1979).

To understand this definition, it's best to first note the difference between MPI and simple medical disorders—which are usually produced by identifiable, non-psychogenic causes (e.g., a blow to the head, a poisonous gas, or a flu virus). It's also useful to note the difference between MPI and traditionally defined psychosomatic illnesses such as ulcers, high blood pressure, and asthma—in which there may be a large psychological component but whose effects appear more gradually and involve more obvious damage to the body.

To illustrate how the term MPI is applied to the workplace, I will cite four examples. The first two are often-cited classics, while the others are less well known, include an obvious case of mislabelling, and are more representative of what has become known as assembly-line hysteria.

Classic Examples

Probably the best known case of work-related MPI is described by the sociologists Kerckhoff and Back in their book *The June Bug* (1968). It was reported as a case of "hysterical contagion" involving approximately sixty textile workers in Spartanburg, South Carolina in the summer of 1962. According to these researchers, to local M.D.s, and to press accounts, over the period of a few days, scores of mostly female workers at one textile plant reported being bitten by insects while working, and then suffered headaches, shortness of breath, severe nausea, numbness, fainting, and convulsions. Some workers were hospitalized, the rest were treated and sent home, and the plant was sprayed for bugs. Since only one chigger could be found in the plant, since no hazardous dyes were isolated, since many of the victims seemed anxious, and since the illness spread so rapidly, it was attributed to chiefly psychological causes. It was characterized by Champion & Taylor (1963) in the *Journal of the South Carolina Medical Association* as "mass hysteria associated with insect bites."

In analyzing the outbreak of this illness, Kerckhoff and Back noted how a number of social variables, such as friendship patterns, were related to the spread of the disorder. In general, they concluded that it could be treated as a type of social contagion triggered by family and work pressures, such as significant amounts of forced overtime (Kerckhoff, Back, & Miller, 1965).

A more recent, but increasingly well-known case of MPI occurred in 1972 in the Data Processing Center of the University of Missouri (in Columbia, Missouri), and was described by the sociologists Stahl and Lebedun. According to their reports (Stahl, Lebedun, 1974; Stahl, 1982), a majority of the 35 keypunch operators (all females) working in a single room at the Data Center developed symptoms such as headaches, dizziness, nausea, fainting, and vomiting on two successive days. While the affected workers attributed their illness to an unidentified gas, environmental engineers could find no evidence of toxic gases when they examined the work place.

As described by Stahl and Lebedun (1974), this apparent case of MPI occurred in an extremely stressful work environment: keypunch operators worked individually at old, noisy keypunch machines in one large room, were not allowed to talk to each other except during two 15-minute breaks (to which they were randomly assigned), were constantly watched by a supervisor in a glass-fronted office (which looked down on the work room), and had to obey their (male) supervisor's arbitrary dress code. At the time of the incident of MPI, these workers had been subjected to weeks of heavy construction noise (including dynamite blasts with no warning) from an adjacent site, and also a large diesel engine (for running air-hammers) had recently been moved under an open window of their work room. Finally, all the keypunch operators were working a large amount of forced overtime at the time of the incident.

In analyzing this case of MPI, Stahl & Lebedun (1974) suggested that "an occasion of 'legitimated illness' . . . was resorted to for relief [from work-related stress]. The results of this collective episode were obviously successful, at least temporarily, in relieving the accumulated and continuing strain." Also, these authors point out, "the collective action of the workers [made] them relatively impervious to punishment" (p. 49). Despite this relatively straightforward explanation for the affected workers' behavior, Stahl and Lebedun titled their article ". . . an analysis of mass *hysteria.*"

A More Typical Case

In contrast to these classic cases, most people's understanding of MPI (or mystery illness or assembly-line hysteria) is based on cases which are either less thoroughly investigated or less accurately reported to the public. As a result, probably the majority of cases that non-professionals hear about cannot be reliably categorized as either MPI or not-MPI.

One such case of alleged MPI occurred in Portland, Maine (Wine, Note 2) and (as far as I know) has not been cited in any professional publications. Since most local health records from that time have been destroyed (Hinckley, Note 3), and since recent efforts to interview the affected workers have been unsuccessful, I have had to rely primarily on the same sources that brought this to the attention of the general public—a series of articles in the local newspaper (Clifford, 1972; Ladouceur, 1972a, b; Maraghy, 1972).

On May 18, 1972 it was reported that approximately thirty-five workers at a Portland Health-Tex plant (manufacturing children's clothes) were treated at local hospitals for nausea, numbness, and fainting. At the end of the day before this incident, some workers had smelled diesel fumes in the plant. At 8 a.m. the next day, the outbreak of illness was said to have begun with five workers fainting, shortly after they had arrived for work. In response to this, all employees were evacuated to the parking lot, where about thirty more became faint, short of breath, and nauseous. Although a local air pollution engineer speculated that the plant's fans may have pulled boiler fumes into the work area, air sampling proved negative, and local doctors failed to identify "classic carbon monoxide poisoning symptoms." In the end, the workers' illness was attributed to nervousness, hyperventilation, and hysteria (Kamin, Note 4; State of Maine, Note 5). All the affected employees were women.

As in many cases that are categorized as "psychogenic" or "mystery" illnesses, this example from Portland involves the suspicion of poor ventilation in a non-union shop, with a mostly female work force that is engaged in highly repetitive work, probably under high production pressure. This case also involves the implicit diagnosis of psychogenic illness based on: (1) the belief of local hospital physicians that poisoning from toxic fumes could be ruled out, and (2) local health officials' inability, after the fact, to confirm workers' reports of hazardous conditions on the job.

A Clear Case of Industrial Poisoning

Because of the difficulties involved in fully investigating and reporting cases of suspected MPI, each year there are probably scores of similar, small-scale industrial epidemics and temporary plant evacuations—whose causes remain officially undetermined. At the same time, however, there is probably a larger number of cases in which epidemic illness in a plant is successfully traced to the effects of a known toxic substance. The relevance of such cases to the understanding of MPI is that many cases of industrial poisoning are labeled as MPI by management, in an attempt to blame workers for getting sick.

To cite just one example that occurred in Connecticut in October 1978, a group of workers at National Semiconductor's Danbury plant were hospitalized as the result of what turned out to be a leak of potentially lethal Diborane gas. Although an outside consultant was eventually able to discover the leak and to relate the workers' illness to its effects, management initially tried to pass the event off as a case of "hysteria" (Hurley, 1978a). Not only was this shown to be untrue, but it turned out that the affected plant had a history of safety violations and a reputation for disregarding previous employee complaints about unsafe working conditions. For example, in just the two months before the Diborane leak, the plant had been twice investigated by the Occupational Safety and Health Administration (OSHA), and cited for a wide variety of health and safety dangers. Such hazards had resulted in an average of one employee per week going to the hospital with a work-related injury.

At this affected plant, when employees developed symptoms

of toxic poisoning and complained to supervisors, they were usually told that their problems were "all in their head" (Hurley, 1978b). Thus, even when twelve workers were taken to the hospital and the plant was evacuated as a result of the Diborane gas leak, production was resumed the same evening without the leak being fixed. As a result, at least six night-shift workers became ill, had to be taken to the hospital, and the plant was again evacuated ("Mystery fumes . . .," 1972).

According to the consultant who later identified the source of the workers' illness, such incidents are typical. He noted, "It could happen, and is happening in industries all over the country. Generally, these incidents are a reflection of poor ventilation, gas leaks, and a bad handling of toxic chemicals ("Illness at National . . .," 1978). In the experience of another health official, "It occurs more often than not, but generally there's no publicity—the company brings in a maintenance man to fix that particular problem, and work continues" ("Illness at National . . .," 1978). Of course, by that time workers may have already become disabled or permanently sensitized to certain chemicals (Martinez & Ramo, 1980).

The Scientists' Explanations

Incidents of apparent mass psychogenic illness, such as the first three cases described above, have become an increasingly popular topic for researchers in epidemiology, medicine, psychology and sociology. For this reason alone, these professionals' explanations for MPI at the workplace deserve careful examination.

One possible explanation for such events is the psychiatric concept of hysteria, which initially signified a wandering uterus that disturbed the functioning of otherwise normal parts of the body. That, of course, was why women were the only ones who could qualify for the diagnosis of hysteria.

My impression is that few industrial psychologists or psychiatrists in the United States would today see industrial "mystery illness" as simple cases of hysteria, even according to the modern (e.g., less intrapsychic) equivalents of that concept. They would be more likely to agree with explanations like those proposed by the psychologist Michael Colligan, who is the most widely published specialist in the field and currently a research psychologist at the Behavioral and Motivational Factors Branch of the National Institute for Occupational Safety and Health (NIOSH). In reviewing sixteen reported incidents of "contagious psychogenic illness" in schools and work places, Colligan attributes these incidents to five basic factors related to the work environment (Colligan & Murphy, 1979). These factors are:

(1) workers' boredom with mindless, repetitive jobs;
(2) unreasonable production pressure;
(3) "physical stressors," such as the excessive noise in the Data Center case;
(4) unresolved complaints against work supervisors and management; and
(5) a lack of social support for workers (e.g., no chance to talk to co-workers on the job).

According to this formulation, an outbreak of MPI can occur when one or two workers come down with an identifiable problem (e.g., nausea) due to an identifiable cause (e.g., an insect bite, a noxious odor), triggering the release of pent-up stress, frustration, and anxiety in other workers. Supposedly, this release takes the form of anxiety-related problems such as shortness of breath, nausea, headaches, and fainting.

In answer to the question of why women seem to develop these contagious symptoms at work much more than men, psychologists generally offer one of two different explanations. The more conservative, deterministic view is that women's greater passivity and suggestibility makes them protest in an indirect way, using the relatively passive "vocabulary" of physical symptoms (Bart, 1968; Singer, Note 6). The more liberal view is that women are forced into the most stressful, most often non-unionized, lowest paid work, and are subject to social pressures to perform full-time homemaker chores at the same time that they're working a forty or fifty hour-a-week job. Hence it is women that are under more stress and are more likely to be "triggered" into acute, stress-related symptoms (Colligan & Stockton, 1978).

Problems with these Explanations

What is wrong with even these relatively liberal views of what causes "mystery illnesses" on the job? Don't explanations such as Colligan's take into account most of the relevant causes of MPI? Aren't these theories also fairly progressive, since they focus on the work environment as the basic source of these disorders (for contrast, see Maguire, 1978)?

I would argue that the social/psychological theories outlined above are seriously incomplete. That is because these theories, and their associated research programs, fail to adequately answer some basic questions related to health, work, and society. In the remainder of this chapter I will consider three such questions. The first is: How do the news media (and indirectly, big business) present the concept of mystery illness or MPI? The second question is: What actually takes place in industries, plants, and shops where MPI is said to break out? My third question is: What roles do individual scientists and agencies such as NIOSH play in regard to reported cases of MPI?

The Media and Mass Psychogenic Illness

Documenting the media's view of the problem of MPI is difficult. First, radio and television accounts of suspected cases of MPI are difficult to find, even when one can pinpoint the time and place of such a case using other sources. Second, even if one takes newspaper reports as representative of mass media accounts in general, most cases of suspected or inferred MPI are never reported in newspapers with wide circulation or published indices. Nevertheless, it is possible to find a few cases in which local news coverage can be examined.

In my opinion, local newspaper coverage of MPI often suffers from two types of errors. First, local papers usually do not cover the day to day problem of non life-threatening, but potentially debilitating health hazards on many jobs. Occasionally, they will publish an article on an unexplained disorder affecting groups of workers. However, because little background is provided for such a story the reader is left with the impression that the major threat to workers' health today is from sources which are uncontrollable, infrequent in their effects, and probably psychogenic in nature.

The second problem with local press coverage is that reporters are frequently forced to rely on plant management or on pro-management health officials for information. Most often, these sources attribute any collective health problems to mass hysteria, although this is usually said off the record, and reported as "one possible cause" being discussed by "officials," "investigators," or "others" (Hurley, 1978; McQuaid, 1979; Stuart, 1976). Coupled with reporters' frequent emphasis on the affected workers' nervousness and on the fact that they happen to be women (Clifford, 1972). This type of innuendo can result in the same biasing effect as the headline: "Mystery Bug Hits Girls at Factory" (McEvedy, 1982).

The View of National Newspapers

Turning from local papers to national dailies such as *The New York Times* and *The Wall Street Journal*, one generally finds reporters writing with more apparent detachment. At the same time, underneath these national papers' more sophisticated coverage one still finds an anti-labor bias that affects the reporting of psychological and health-related issues.

The New York Times' coverage of the topic MPI is an interesting example of such bias. On one level the *Times* seems to want to set the record straight on MPI, giving full coverage to the liberal theories of Colligan and to his warning that research on MPI might be used by some companies to screen out women and minority workers—who would supposedly be more likely to come down with a stress-related disorder (Peterson, 1979). However, at the same time that it provides this liberal warning about MPI, the *Times* consistently avoids covering the day-to-day health and safety problems of workers in the New York area. As a result, when the *Times* prints an article with the title "No Remedy for Illness at L.I. [Long Island] Plant" (Silver, 1979), about women clerical workers getting sick from what turned out to be a bad ventilation system, the *overall* impression is that the biggest problem facing workers is "mystery illnesses"—which, if everyone's lucky and patient, might be solved by outside experts.

In contrast to the *Times*, the *Wall Street Journal* is more direct in revealing its biases. For example, in March 1980 the *Journal* published an article about so-called mystery illness among electronics workers in Malaysia that is openly racist and sexist in its description of Malaysian workers (Newman, 1980). It's titled "Malaysian malady: When the spirit hits . . .," and subtitled "Spooks stir up mass hysteria in women in factories; a way of raising the pay?"

In addition to showing how much obvious racism the *Journal* is willing to exhibit, this article reveals three interesting facts about MPI, the press, and U.S. multinational firms. First, the article's author admits that he is chiefly interested in outbreaks of MPI because of their lowering of the output of factories owned by U.S. multinational firms. He says:

> Nearly all the producers of silicon nerves for electronic brains are here [in Malaysia]: companies such as Texas Instruments, RCA, Harris Semiconductor and Motorola each employ up to 6,000 dexterous young women. But every so often, one of these ordinarily docile Malays will spy an apparition in her microscope and give her factory the fits. . . . To let the dust settle, factories have to be closed for a week or more (p. 1).

The second fact that can be found in this article is the reason for these young women workers' succumbing to the stress of their jobs. That is their being paid as little as $1.50 per day, having no union, no medical benefits, no grievance procedure, and no maximum work day; all they have are company spies and goons who "drag a victim from the [shop] floor before her delirium becomes infectious" (p. 37).

A final point that can be seen by a close reading of this article in the *Journal* is the ambivalence that company managers feel toward industrial psychiatrists and psychologists. Although these consultants are paid to show managers "how to handle hysterical factory workers" (Chew, 1978). The methods of control which the consultants recommend are more indirect and less confrontative than many managers normally use. Thus, when the *Journal* article describes a plant manager's mixed feelings about having to call in a native spiritualist/healer to calm a group of workers in Malaysia, it sounds very much like a parable about a manager in the United States calling in a psychiatric consultant.

If my perception of this last point is correct, it also suggests that there is something quite threatening about these industrial epidemics, whether psychologically caused or not. How else could they induce management to use services and techniques for which they have traditionally had such mistrust? In my view, some of the threatening quality that comes through the various press accounts is the collective nature of the episodes, and their inherently subversive quality. That is, management finds it much easier to handle an individual worker who rebels openly than to cope with a group of workers who shut down production while saying that they *want* to work, not to be sick.

Hazards at Work

The second question that is rarely addressed in the literature on mass psychogenic illness is: "What are the actual working conditions in the places where MPI is diagnosed?"

Although a complete answer to this question will require the analysis of additional cases, (Garrett & Forman, Note 7), I can begin here by focusing on the electronics industry. In discussing the *Wall Street Journal*'s article on Malaysia, I've cited a few of the oppressive working conditions that are maintained by U.S. multinational corporations abroad. Because of space limitations I must refer the reader to a recent article for a more complete discussion of this subject (Ehrenreich & Fuentes, 1981). To cite just one detail from this article that is never mentioned in the psychological literature, a typical electronics production job in South Korea or Malaysia is held by a woman under thirty years old, pays about $2 per day, involves peering through a microscope at electronic components for seven to nine hours each day, and results in a vast majority of workers developing severe eye problems *within the first year of their employment*. Psychologists may wonder why Malaysian electronics workers begin to see apparitions in their microscopes; apparently, after a few years of such work, they're lucky to see anything at all.

Turning to the United States, electronics is on NIOSH's list of high health-risk industries, and in areas such as California's "Silicon Valley," death or serious illness from toxic chemicals at work is a frequent occurrence (Thurlow & Cooney, 1976). To those who suspect that labels such as MPI may cover up toxic poisoning, it is no surprise that so many examples of mystery illness are now being reported in the electronics industry.

For example, in the fall of 1976, well publicized cases of mystery illness were reported in six different states, all at firms making electronic components ("Mystery Ailments," 1976). Although some health officials suggested that workers' symptoms were mostly psychogenic, an OSHA investigation eventually found them caused by "poor ventilation, coupled with the use of [certain] solvents and chemicals in the workplace" ("Ventilation Blamed," p. 23). Thus, at least from press reports, these six cases sound similar to the Danbury, Connecticut, incident mentioned earlier—a case of a hazardous job and management indifference to workers' complaints.

Given these examples of apparent MPI that, upon careful investigation turn out to be industrial poisoning or unbearable work pressure, one wonders about the industrial hazards that may have contributed to other, more famous cases of MPI. While I do not expect that all researchers will come to share my suspicions about each shop in which MPI is reported, I do suggest that industrial work places, particularly nonunionized ones and ones in hazardous industries such as electronics, should be considered hazardous until fully investigated and described in published reports.

The Scientist and the Diagnosis of MPI

Why am I so suspicious of health conditions at work places? Why do I seem to doubt the accuracy of many reports of MPI, even those appearing in professional journals? To answer these questions, I will turn to a consideration of the scientist's role in this field—first the scientist in the role of investigator and diagnostician of MPI, and then the scientist as someone who operates within a political and institutional context.

In investigating a case of MPI there is a fairly standard procedure that is followed. First, environmental tests must be made to see whether significant concentrations of toxic agents are present. Then, workers who have become ill must be identified, interviewed, and given psychological and sociological tests. For a diagnosis of MPI to be made with confidence, not only must environmental toxins be ruled out, but also the victims of MPI should possess some common personality characteristics (e.g., a high degree of suggestibility), show relatively similar physical symptoms, and be related to each other on some sociological measures (e.g., friendship choices).

Although this may sound like a reasonable procedure, each of its steps can be criticised on logical or empirical grounds. For example, as Boruch (1982) has explained, most accounts of MPI do not provide enough information for the reader to judge whether the environmental tests were done systematically enough to be valid, or whether the data collected are reliable indicators of environmental conditions. Also, since one can never logically *prove* the negative case (i.e., that a toxin isn't present), environmental tests can only *suggest* that there is a psychological cause for an incident of industrial illness (McGrath, 1982). Such causality cannot be proven, strictly speaking.

Even if one were able to rule out toxic poisoning, other problems with MPI diagnostics soon appear. First, separating affected from non-affected workers is an inherently unreliable procedure, since the severity of symptoms varies between workers (so some may not be reported), and since symptom assessment is usually done after a factory has been evacuated and the workers have been dispersed (McGrath, 1982). Also, data from the interviewing and the sociological/psychological testing is likely to be distorted by either experimenter bias, the passage of time, or by the subject's experience after the initial outbreak of illness (Boruch, 1982). Finally, even if personality trait differences are found between workers following a particular incident, these differences may fail to generalize to other situations, and may be useless in the prediction of who will be a victim of MPI in the future (Tornatsky, 1982).

Given all these problems inherent in the search for psychological factors that underlie cases of suspected MPI, one would think that such a search might be called off—at least until some laboratory research showed that valid tests of MPI-related traits could be designed and put to use (McGrath, 1982).

The Scientist Within an Institutional Context

Despite some strong criticism of the MPI literature from respected sources, it is unlikely that the direction of this research can be easily changed. A major reason for this is the limitation imposed on the individual researcher by his or her institutional role. The more that I look into this problem, the *less* I see the study of MPI as another example of the supposedly natural progress of science toward some objective, impartial truth. To cite just one obvious constraint on the validity of the research in the field, in the vast majority of the cases which I've studied, the individual researcher was never free of the company's control over some aspect of the investigation.

This problem of management interference in health investigations is most severe in largely unorganized sectors of industry, such as the manufacture of electronic components. In Sunnyvale, California, for example, a recent case of suspected industrial poisoning illustrates the constraints on those who investigate industrial hazards—whether they are in the role of psychologist, sociologist, physician, or epidemiologist (Bacon, Note 8).

In this case, three women employees of the Signetics Corporation complained to state and federal health agencies in March 1979 about a variety of debilitating symptoms (e.g., burning tongue, mouth sores, headaches, tachycardia) that they attributed to the long-term use of certain chemicals in their work. Despite their complaints to management over the previous two years, the plant had not been cleaned up, and these women were still unable to enter many work areas in the plant without becoming ill. As a result of the workers' petitioning, NIOSH and Cal-OSHA consultants conducted a series of site visits to the Signetics plant. Based on plant records and data collected during these visits, the consultants concluded that a "significant occupationally-related health problem" existed at the plant (Belanger, Note 9). At the same time, however, the consultants were never given full access to the plant, to health records, or to workers (Nelson, 1980). Thus, the consultants were unable to make a definitive report on the overall health problems at Signetics, or to recommend specific changes in the plant to protect the majority of the workers. Also, during the course of the investigation, it became clear that the three most affected workers were permanently sensitized to certain chemicals. These workers were fired by Signetics in July, 1979.

The Signetics case not only shows management interference with the work of individual researchers, but there is some evidence that this is also an example of industry's attempted influence on the regulatory charges that NIOSH's report of hazards at Signetics has prompted local electronics companies to begin lobbying to have NIOSH and Cal-OSHA be more charitable in their interpretations of data found at such plants. Even if this lobbying turns out to be unsuccessful, agencies such as NIOSH are generally at the mercy of the industry, since, for example, the safety of most of the chemicals used in electronics has not been fully tested—either individually or in combinations with potentially synergistic effects (Berman, 1979; Case, 1980). As a result, the setting of many safe dosage levels is based primarily on industry recommendations. Given this technological/institutional context, and given the federal government's current attacks on the independence of OSHA (Bedell, 1981; Peterson, 1981), it seems unlikely that the reliability and validity of research in this area will improve in the near future.

What Can be Done?

I have tried to describe the anti-labor, anti-woman ideology that surrounds concepts such as MPI assembly-line hysteria. I will conclude by sketching some possible ways to help prevent the further victimization of working women by those who put this ideology into practice.

First I will mention without comment an important, but difficult task: strengthening NIOSH and OSHA, including making them more independent of industry. Next are the related goals of organizing the unorganized majority of women in industry and strengthening rank and file movements within unions—such as local women's committees. While these are not tasks which can be accomplished primarily by academic researchers or health professionals, I mention them here as a reminder of the connection between women's health on the job and the general structuring and control of work itself (Berman, 1979). To ignore this connection would make it even more difficult to develop a clear understanding of the problem of MPI, or of what can be done about it.

A third possible antidote to reactionary uses of the concept of MPI is the creation of a more responsible professional literature on this subject. For example, journal articles could be written that emphasize the fact that not all victims of industrial psychogenic illness are women (Delasnerie, 1972/73)—a topic that is neglected today. Also, the literature could begin to reflect more accurately important job-related variables at MPI-affected plants, such as the amount of overtime worked, the plant's history of labor relations (e.g., is there a union?), its wage incentives for workers (e.g., piece work rates), and the plant's health and safety record.

More basically, I believe that the literature's sample of MPI cases should be enlarged to include the high frequency of false positives (pseudo-MPI cases) that are reported each year. In a sense, these cases are failures to replicate the previously reported connections between apparently psychogenic symptoms (headache, nausea, fainting, nervousness) and the absence of toxic poisoning. Instead of ignoring these examples of non-replication, the literature should regularly contain descriptions of them. Such descriptions would serve as a reminder of the problems involved in diagnosing MPI, and of its unreliability as a psychological concept.

Finally, in my list of recommended actions, I would urge that the term "mass psychogenic illness" *not* be made a "legitimate diagnostic category" within medicine or psychiatry, as some have suggested (Colligan & Smith, 1978). Rather than emphasizing the *psychogenic* nature of what gets called MPI, academic researchers and health officials should place more emphasis on the *environmental* causes of these workers' becoming ill.

If a label for such disorders were still necessary, I would recommend calling them something like "environmental stress reaction" or "industrial toxin syndrome." Although these terms may be little better than others already proposed, the selection of labels is of secondary importance to me. More crucial is how such terms are used. Thus, I would recommend saving *any* labels in this area for use only in those contexts where they're not easily abused (e.g., as a journal index term). For other occasions (e.g., newpaper interviews), I suggest an emphasis on stressful working conditions rather than on diagnosing what type of "illness" each worker is said to have.

I do not present these recommendations as an expert who hands down a completely developed solution to the problem of industrial psychogenic illness. Rather, my intention is to encourage further study and action on this complex topic. Only through such activity, I believe, can we hope to ensure that the label of assembly-line hysteria will cease to be used against specific groups of working women, and against women industrial workers in general.

FOOTNOTE

Research for this paper was begun as part of the author's participation in an NSF/Chautauqua Course: *Culture, Body, and Behavior,* taught by Ruth Hubbard and Marian Lowe. Bibliographic assistance was provided by James Forro, and by Shirley Maul and the staff of the Vassar College Library. I thank Jagna Scharff, Don Kamin, Mary-Agnes Wine, James Melius, William Allen, David Bacon, and the Project on Health and Safety in Electronics for contributing information on individual cases of mystery illness. Thanks also go to Michael Colligan for his generous cooperation, to Becky Mitchell for helping edit a draft of this paper, and to Terri Cronk for her secretarial help. Finally, I am grateful for the support of the organizers and participants in the Fourth Conference on Genes and Gender.

NOTES

[1] National Institute for Occupational Safety and Health. *Symposium on the diagnosis and amelioration of mass psychogenic illness.* Chicago, Illinois May 30–June 1, 1979.

[2] Wine, Mary-Agnes. *Personal communication to author.* October 24, 1980.

[3] Hinckley, W. W. *Personal communication to author.* February 11, 1981.

[4] Kamin, D. *An investigation of reported mass psychogenic illness.* Paper presented at the annual meeting of the New England Psychological Association, Brandeis University, October 31, 1981.

[5] State of Maine. Department of Environmental Protection. [*Report on inspection of Health-Tex Plant*]. Augusta, Maine, 1972.

[6] Singer, J. E. *Yes Virginia, there really is a mass psychogenic illness.* Paper presented to NIOSH Conference on Mass Psychogenic Illness, Chicago, May/June, 1979.

[7] Garrett, K. & Forman, K. *An investigation of a case of mystery illness at the University of Massachusetts.* Unpublished manuscript, Vassar College, 1981.

[8] Bacon, D. *Personal communication to author.* January 26, 1981.

[9] Belanger, P. L. *Interim report #2, Health hazard evaluation project #HHE 79-66.* [prepared for the] National Institute for Occupational Safety and Health, Cincinnati, Ohio. January 31, 1980.

REFERENCES

Bart, P. B.
    1968   Social structure and vocabularies of discomfort; What happened to female hysteria? Journal of Health and Social Behavior, 9, 188–193.

Bedell, B.
    1981   Critic of job safety heads OSHA. The Guardian.
    Feb.

Berman, D. M.
    1979   *Death on the job: Occupational health and safety struggles in the United States.* New York Monthly Review Press.

Boruch, R. F.
1982 Evidence and inference in research on mass psychogenic illness. In Colligan, M., Pennebaker, J. W., and Murphy, C. (Eds.), *Occupational health and social behavior: Mass psychogenic illness.* Hillsdale, N.J. Erlbaum.

Case, L.
1980 Mar./April Mass psychogenic illness. Science for the People, pp. 18–19.

Champion, F. P., Taylor, R., Joseph, P. R. & Hedden, J. C.
1963 Mass hysteria associated with insect bites. Journal of the South Carolina Medical Association, 59, 351–53.

Chew, P. K.
1978 How to handle hysterical factory workers. Occupational Health and Safety, 47, 50–52.

Clifford, K.
1972 May 19 Malaise at local plant 'like end of the earth.' Portland, Me. Press Herald.

Colligan, M. J. & Murphy, L.
1979 Mass psychogenic illness in Organizations: An overview. Journal of Occupational Psychology, 52, 77–90.

Colligan, M. J. & Smith, M. J.
1978 A methodological approach for evaluating outbreaks of mass psychogenic illness in industry. Journal of Occupational Medicine, 20, 401–402.

Colligan, M. J. & Stockton, W.
1978 June The mystery of assembly-line hysteria. Psychology Today, pp. 93–99; 114–116.

Delasnerie, R.
1972/73 Conduite hystérique parmi le personnel masculin dans l'industrie. Psychotherapy and Psychosomatics 21, 149–152.

Ehrenreich, B., Fuentes, A.
1981 Jan. Women and multinationals: Life on the global assembly-line. Ms., pp. 52–59; 71.

Hurley, J.
1978 Oct. 19 OSHA joins probe of factory fumes. News-Times Danbury, Conn., p. 1. (a)

Hurley, J.
1978 Oct. 24 Safety violations riddle city plant. News-Times Danbury, Conn., pp. 1; 8. (b)

Illness at National traced to gas leak. Electronics, pp. 42; 44.
1978 Nov. 9

Kerckhoff, A. C., & Back, K. W.
1968 *The june bug: A study of hysterical contagion.* New York Appleton-Century Crofts.

Kerckhoff, A. S., Back, K. W., & Miller, N.
   1965   Sociometric patterns in hysterical contagion. Sociometry 28, 2–15.

La Douceur, G.
   1972   Air pollution is ruled out in mystery malaise here. Press Herald Portland, Me., p. 14. (a)
   June 6

LaDouceur, G.
   1972   Officials theories differ on illness at local plant. Press Herald Portland Me. (b)
   May 19

Maguire, A.
   1978   Psychic possession among industrial workers. Lancet, pp. 367–377.
   Feb. 18

Maraghy, G.
   1972   33 persons are felled at plant here. Press Herald Portland Me. p. 1.
   May 18

Markush, R. E.
   1982   Trends in mental epidemic diagnosis and investigation. In Colligan, M., Pennebaker, J. W., & Murphy, C. (Eds.), *Occupational health and social behavior: Mass psychogenic illness*. Hillsdale, N.J. Erlbaum.

Martinez, S., & Ramo, A.
   1980   In the valley of the shadow of death. In These Times, pp. 12–13.
   Oct. 8–14

McEvedy, C. P.
   1982   An epidemic in a textile plant in the English midlands. In Colligan, M., Pennebaker, J. W., Murphy, C. (Eds.), *Occupational health and social behavior: Mass psychogenic illness*. Hillsdale, N.J. Erlbaum.

McGrath, J. E.
   1982   Complexities, cautions and concepts in research on mass psychogenic illness. In Colligan, M., Pennebaker, J. W., & Murphy, C. (Eds.), *Occupational health and social behavior: Mass psychogenic illness*. Hillsdale, N.J. Erlbaum.

McQuaid, E. P.
   1979   Source of sickness at research center eludes experts. Daily Hampshire Gazette, pp. 1; 11.
   May 19

'Mystery' ailments at electronic plants are studied for link. Wall Street Journal, p. 7.
   1976
   Oct. 8

Mystery fumes close city plant. News-Times, Danbury, Conn. pp. 1; 8.
   1978
   Oct. 18

Nelson, Mark.
   1980   Federal team probes health risks at Signetics. Valley Journal Cupertino, Calif. pp. 1; 8.
   Feb. 20

Newman, B.
1980 Mar. 7   Malaysian malady: When the spirit hits, a scapegoat suffers. Wall Street Journal pp. 1; 37.

Peterson, C.
1981 Feb. 12   Reagan fills three sub-cabinet posts. Washington Post p. A3

Peterson, I.
1979 May 29   Stress can cause work epidemics. New York Times pp. C1–C2.

Silver, R. R.
1979 Sept. 12   No remedy for illness at Long Island plant. New York Times, p. B2

Sirois, F.
1975   A propos de la fréquence des épidémies d'hystéria. L'Union Medicale du Canada, 104, 121–123.

Stahl, S. M.
1982   Illness as an emergent norm, or doing what comes naturally. In Colligan, M., Pennebaker, J. W., & Murphy, C. (Eds.), *Occupational health and social behavior: Mass psychogenic illness.* Hillsdale, N. J. Erlbaum.

Stahl, S. M. & Lebedun, M.
1974   Mystery gas: An analysis of mass hysteria. Journal of Health and Social Behavior 15, 44–50.

Stuart, R.
1976 Nov. 16   Plant in sixth state has mystery illness. New York Times, p. 54.

Thurlow, G., & Cooney, P.
1976 Nov. 19   Death in building G. Santa Barbara News & Review

Thornatzky, L. G.
1982   Issues in the implementation of remedial interventions in psychogenic illness. In Colligan, M., Pennebaker, J. W., & Murphy, C. (Eds.), *Occupational health and social behavior: Mass psychogenic illness.* Hillsdale, N.J. Erlbaum

Ventilation blamed
1977 April 20   Chemical Week, p. 23.

GENES AND GENDER IN THE WORKPLACE

Judith S. Bellin, Ph.D. and Reva Rubenstein, Ph.D.

With some notable exceptions, the differences in susceptibilities to the toxic effects of chemicals on male and female animals of the same species usually are slight: Age, weight, and developmental maturity are far more important determinants of toxicity than is sex. (Doull, 1979)

Toxic chemicals do not discriminate—they affect both female and male hearts, muscles, livers and kidneys—and both female and male reproductive systems. However, because their reproductive systems differ, there are differences in the ways in which toxic chemicals affect the reproductive function of men and women. Regrettably, these differences are used to deny equality of opportunity—we must therefore clearly examine and define what is known, so that we may consider how to take the differences into consideration without violating equality of opportunity.

Women have worked in situations similar to men for many generations. Because of the prenatal growth and development of the young, women and their offspring have thus been exposed to the many toxic agents of the workplace for many years—why then the new outcry about the susceptibility of women and their offspring to the noxious agents of the workplace? How realistic is this concern? Are women and the fetus at special risk in the workplace? If so, do they need special protection? Do such special needs, if any, counter the mandates of the Equal Employment Opportunity Act? We must examine and define what is known, so that we may develop answers to questions such as these.

This discussion is limited to a consideration of what is known about sex-based genetic differences in reaction to toxic chemicals. Consideration of the facts regarding other genetic susceptibilities is equally important, and deserves a full treatment. In order to give perspective to our concern, another

HUMAN REPRODUCTION(3)

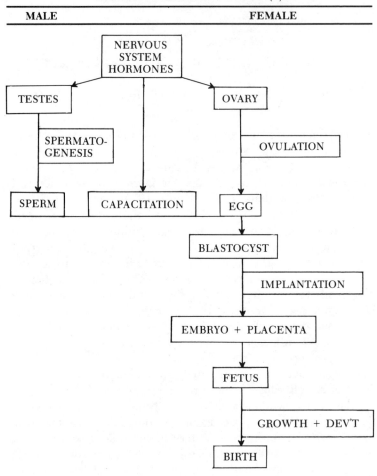

FIGURE 1: Some salient features of human reproduction.

contributor on this program describes the present demographics of women's participation in the industrial workplace.

Reproductive systems consist of germ cells (sperm or eggs), the specific organs that produce them, and a series of specialized tubes and their associated glands, which convey the gametes to each other for fertilization. In higher animals, one of these specialized tubes, the uterus, is the site of implantation of the fertilized egg (the blastocyst). (Vander, 1975) The prostate and seminal vesicles secrete substances which assure the nutrition and well-being of the sperm. These secretions constitute a possible route of chemical interference, since chemicals in the blood may be carried into the secretions of the glands. (Council on Environmental Quality, 1981)

In both males and females a group of cells synthesize the sex hormones necessary to keep the reproductive system functioning properly. These hormones are themselves regulated by centers in the brain, the hypothalamus and the pituitary.

There are important points of difference between the male and female reproductive systems. A very important point of difference is the fact that only female mammals enable the development of the embryo within the uterus until birth. Eggs and sperm differ in maturation rates. In the female, growth of eggs ceases during embryonic development. A female is born with the full complement of eggs she will have during her life. This number of human eggs (approximately 35,000) gradually undergoes absorption (atresia) throughout life, and only about 400 reach maturity. Of these, only a small fraction are fertilized. The older the egg cells are, the more environmental insult they are likely to have sustained, and therefore the rate and quality of reproductive outcome decreases as women get older. (Vander, 1975; Odell, 1971)

In contrast, the male germ cells continue to be produced from puberty until late in life. Because they are capable of replenishment, insult to these cells may be reversible, depending on the extent of the damage. Indeed, male infertility induced by exposure to dibromochloropropane (DBCP) (Whorton, 1971) and lead (DeKnut, 1973) was shown to be reversible when exposure ceased. Both eggs and male germ cells are susceptible to damage from viral infection (leading to male sterility from mumps, for example), to radiation, and to the action of chemicals and drugs.

Chemicals may modify a cell's genetic material, its membranes or other structures, its response to hormones, or its metabolic pathways.

Nature has arranged some defenses: the blood/testes barrier excludes some chemicals from reaching the sperm. However, recent evidence indicates that this barrier is not very effective—especially not with respect to classes of molecules which are lipophilic (fat soluble). The genetic material of sperm cells is relatively unprotected, and it has been speculated (Council on Environmental Quality, 1981) that the affected sperm may be morphologically or functionally incomplete, or may carry genetic defects. It is possible that toxic agents penetrating the blood/testes barrier could act on the Sertoli cells responsible for sperm maturation, on the cells of the accessory glands that produce seminal fluids, and on the cells that synthesize the testosterone necessary for the maintenance of all parts of the male reproductive system. All these effects could produce sperm less able to successfully fertilize an egg.

External agents could also exert an indirect effect on male reproduction by affecting the adrenals, the pituitary gland or the hypothalamus, indirectly affecting the male reproductive process by affecting sperm production or libido. Yet another way in which male reproduction could be affected would be by contamination of seminal fluid with a toxic agent, affecting sperm motility, fertilization or the fertilized egg. It must be emphasized that most of the above mechanisms are speculative—experimental evidence for most is lacking.

In the female reproductive system there are many more points at which, one can speculate, external agents could act. An agent acting at the time of egg formation (remember that this takes place during fetal life) could change the genetic structure of these cells. The many thousands of known mutagens could act in this manner. (Council on Environmental Quality, 1981) This could be manifested as cancer developing when the child grows to adulthood.

Interference with fertilization and reproduction can occur in many ways. A toxic substance may lead to fewer implantations, to an increase in the frequency of resorptions, it can cause the death of the embryo (increasing the frequency of spontaneous abortions), or the abnormal development of its organs. (Council on Environmental Quality, 1981) The results of abnormal development of embryo or fetus, if not sufficient to cause its death, can be manifested in congenital malformations (the agents causing such effects are called teratogens). These malformations may be obvious, as the missing limbs caused by thalidomide, or subtle, as those behavioral and learning deficits caused by *in utero* exposure to lead. (Council on Environmental Quality, 1981) Some agents, particularly hormones, may alter the development of the reproductive system of the fetus (Gill, 1977; Bibbo, 1977), and may result in its impaired development or fertility in adult life. Other toxins, transplacental carcinogens, may initiate a process in the fetus which leads to cancer when it is an adult. (Council on Environmental Quality, 1981) Toxins which act during the period of fetal growth may produce fetal death (stillbirths, spontaneous abortions) or may result in smaller (reduced birth weight), less healthy children. The affected children may be more susceptible than healthy ones to infections, they may develop more slowly than the norm, or have other subtle developmental problems. (Council on Environmental Quality, 1981) (Sullivan, 1979) (USDHEW 1977)

The foregoing discussion is a very abbreviated and superficial summary of the many and subtle ways in which toxic chemicals can affect male and female reproductive function. The interested reader is referred to textbooks on reproductive physiology and toxicology (Doull, 1979; Amelar, 1977; Hogarth, 1978) for more extensive treatment.

Having taken a very superficial look at male and female reproductive physiology, let us now look at some of the data on the effects of toxic chemicals on these functions. Reproductive toxicology is in its infancy. Very little work has been done on the toxic effects to the mechanisms of reproduction of either sex. It had usually been assumed that only the female reproductive role would be of importance in assuring or maintaining reproductive integrity. In older studies the measured outcome is the number and quality of liveborn offspring following dosing of the pregnant female. Newer studies similarly investigate the effects of toxics on sperm by dosing the male before mating to untreated females.

A major problem is that the numerous stages in human reproduction cannot be studied directly, and there are no animal models. Sperm counts and motility are the only variables that can be conveniently studied directly, and little is as yet known about normal variation in these parameters. Other factors can only be inferred from epidemiologic studies, and such data are frequently inaccurate and difficult to analyze (Council on Environmental Quality, 1981). Not all human reproductive impairments or losses are recorded, and the information that is noted (e.g., on birth or death certificates, or in hospital records) is often inaccurate or incomplete. (Barr, 1979) Reproductive failure due to impotence or infertility is not often recorded, losses during the first two trimesters (spontaneous abortions) are not often detected or, if detected, are unrecorded except anecdotally (Council on Environmental Quality, 1981; Barr, 1979). Congenital malformations and the subtle adverse reproductive effects (e.g., biochemical, behavioral or immunologic defects) are not easily detected, and, where detected, are not systematically recorded.

A large proportion of adverse reproductive outcomes, therefore, go unnoticed. Often the realization of adverse reproductive problems is first noticed by the persons affected. Such was the case, for instance, when the sterilant effects of the pesticide DBCP were noticed by the affected workers (Scott, 1978). Much of what is known was discovered in the course of investigations on the effects of drugs used for other purposes, or on public health problems such as abuse of alcohol, addictive or illicit drugs, or smoking.

The data discussed below constitute the sum total of those substantiated. Table 1 is adapted from a particularly useful set of data published by NIOSH (USDHEW, 1977). These are chemicals for which sufficient animal and/or human data are known to allow conclusions on reproductive effects. Bear in mind that blanks in the last five columns probably imply that no studies have been done. Future studies may well note an effect. It is apparent that these chemicals are toxic to more than one organ or system—they do not only act on reproduction. Further, there seems to be a correlation between mutagenic, carcinogenic and teratogenic effects—this is not surprising, since a common mechanism, interaction with the cell's genetic material, is responsible for the first two phenomena, and some reproductive effects are due to this same cause. Tables 2, 3 and 4 illustrate in greater detail the known adverse effects resulting from workplace exposure to some of these chemicals. Often the same chemicals adversely affect reproduction in both men and women.

There is ample evidence that many chemicals affect reproductive functioning of both women and men. Recent reviews (Council on Environmental Quality, 1981; Hunt, 1979; Strobino, 1978.) conclude that the possible mechanisms causing adverse reproductive effects due to paternal exposure can not

TABLE 1
SOME TOXINS AND THEIR EFFECTS*

	EFFECTS											
	SYSTEMIC, OTHER								REPRODUCTIVE			
	NEUROLOGIC	HEPATIC	RENAL	HEMATOLOGIC	CARDIOVASCULAR	PULMONARY	CARCINOGENIC	MUTAGENIC	TERATOGENIC	RED. FERTILITY	INCR. SPONT. ABORT.	RED. BIRTH WT.
HALOGENATED ANESTHETICS	•	•					•	•	•	•	•	•
BENZENE	•			•			•	•	•			
CADMIUM		•	•	•	•	•	•		•	○		•
CARBON DISULFIDE	•				•				•	•	•	
CARBON MONOXIDE	•				•	•						•
CHLOROPRENE	•	•	•	•			•	•	•	○		
DBCP		•	•				•			○		
DES					•		•	◻	•			
ETHANOL		•			•		•	◻	•		•	
ETHYLENE DIBROMIDE		•	•			•	•			○		
KEPONE	•	•					•			○		
LEAD	•		•	•			•	•	•	○	•	•
MERCURY	•		•						•		•	
PCB's		•		•			•		•	•	•	
IONIZING RADIATION					•		•	•	•	•	•	

*Adapted from References 3, 9 and 16.
•Human and/or animal data.
○Evidence only on male infertility.
◻Conflicting reports.

yet be evaluated, because the evidence of paternal exposure is scant and conflicting, and that further studies evaluating the risks of environmental exposures on all reproductive and post-

TABLE 2
ADVERSE REPRODUCTIVE EFFECTS OF CHEMICALS ON WOMEN (3, 9)

	CHEMICAL	NOTED HUMAN EFFECTS	OCCUPATION
1.	Lead	menstrual disorders; increased stillbirths, spontaneous abortions, neonatal deaths; decreased birth weight; perhaps congenital anomalies	smelter, battery workers
2.	Beryllium	pregnant women at increased risk of fatal beryllium poisoning	manufacutring, processing
3.	Anaesthetic gases	infertility, increased spontaneous abortions, stillbirths, decreased birth weight	OR, ICU personnel
4.	Formaldehyde	menstrual disorders, toxemia and anemia in pregnancy; increased spontaneous abortions; reduced birth weight	manufacture of urea-formaldehyde resins, medical students, mortuary workers, histologists
5.	Carbon disulfide	menstrual disorders; perhaps decreased fertility, increased spontaneous abortions	rayon workers, chemical operators
6.	Laboratory Solvents	increases in rare birth defects, stillbirths, spontaneous abortions, chromosomal changes	laboratory workers. many industrial workers
7.	PCB's	menstrual disorders; birth defects	manufacture of chlorinated hydrocarbons.

natal outcomes needs to be carefully explored.

The many known (and undoubtedly many as yet unknown) reproductive hazards which are potentially harmful to the embryo or fetus are encountered in many workplaces. In addition to the chemical hazards we have listed, we can add the physical agents: radiation (X-rays, radioisotopes, and possibly noise, heat, and vibration), and infectious agents (Hunt, 1979; Stellman, 1978). Some of the occupations in which these agents are encountered are: drug manufacture, work with chemicals and pesticides, hospital work (X-ray, laboratory, operating room, intensive care units, nuclear medicine), battery manufacturing, heavy metal smelting and refining, teaching of small children, animal handling, meat/poultry cutting, industrial and service

TABLE 3
ADVERSE REPRODUCTIVE EFFECTS OF CHEMICALS ON MEN (3, 9)

	CHEMICAL	NOTED HUMAN EFFECTS	OCCUPATION
1.	DBCP	infertility (azoospermia, oligospermia); increased FSH; reversible pathological changes in testes; impotence; chromosomal changes.	manufacture, formulation, use
2.	Lead and/or smelter exposure	infertility chromosomal and sperm abnormalities.	1° or 2° lead smelting, battery mfr.
3.	Kepone	loss of libido; reduced sperm counts; feminization	manufacture, formulation, use
4.	Vinyl chloride	chromosomal abnormalities	manufacture, polymerization
5.	Carbon disulfide	sperm changes, decreased fertility	synthesis, rayon workers

TABLE 4
TERATOGENS ACTING THROUGH MALES

	EXPERIMENT	(spermatogenetic compounds) OBSERVATION	REF.
1.	Male rabbits exposed to thalidomide before breeding	higher death rates in fetal and neonatal offspring	17
2.	Methadone, rats	reduced birth weights, litter size, survival. abnormal behavior on performance tests	17
3.	Lead, rats	reduced litter size, birth weights, survival	3, 17, 18
4.	Anesthetic gases, human	increased spontaneous abortions, congenital defects; higher rate of female births	19, 20
5.	Lead, human	increased miscarriage rate, postnatal mortality; Wilm's tumor in children	1, 3
6.	Vinyl chloride, human	increased spontaneous abortions	22
7.	Kepone, human	reproductive failures	3
8.	Ethylene dibromide, human	decreased fertility of wives	3

work. It has been stated that the fetus is at special risk *i.e.*, that it is more sensititve than adults to the toxic effect of toxic chemicals. (Anon. 1980). This has been proven only in the case of methyl mercury—there is as yet no proof that this is generally true, but the statement is the basis for the exclusionary workplace practices now followed in several industries. (Anon., 1980). All that one can now assert with scientific accuracy is that all living organisms are at risk, differences in the degree of that risk are not generally established. There is however, one important respect in which the fetus is at special risk: those teratogenic defects which are not a consequence of chromosomal change in egg or sperm must be supposed to be the consequence of exposure *in utero* during embryogenesis or fetal development. (Doull, 1979)

What should be done? Should a woman of child-bearing age be excluded from a workplace where her possible fetus could be damaged? Women are not always pregnant: should a policy of "pregnant unless proven otherwise" be adopted?

The revelation, a few years ago, that DBCP causes sterility in male workers caused a fruit grower to remark that it is a good sterilant, and those workers who do not wish children might like to work with it! This crude and distasteful suggestion is, in fact, a not too inaccurate representation of the dilemma confronted by many working men and women.

In many instances management responds to new information about toxic agents which affect reproduction by excluding women of child bearing age from exposure. In 1976 a woman in Canada underwent sterilization in order to preserve her right to work in a GM battery plant (Scott, 1978), and a woman was recently denied the right to work in a vinyl chloride plant—even though the sperm of men employed there is damaged: wives of male workers there have excess miscarriages, and the carcinogenic potential of vinyl chloride affects both men and women. (Scott, 1978.) Although, in such situations, there is usually a paternalistic expression of concern for the woman and the fetus, many suspect that the real concern is for corporate liability. While, in cases of damage to male or female workers corporate liability is limited by state workers' compensation, a child malformed because of toxic exposure could conceivably sue for damages. (Scott, 1978.) (Note, however, that at the present level of scientific knowledge the cause of such malformation would be difficult, if not impossible, to prove.)

Most people work because they have to earn a living. Even

in the absence of clear scientific knowledge employers in the chemical industry act to exclude women of child bearing age—ignoring the accumulating evidence that males are also at reproductive risk, and the fact that reproductive toxins usually have other serious systemic effects as well. Neither unions nor equal opportunity activists have developed a strategy for dealing with such selective employment practices. The industrial hygienist for one union is quoted as being stumped because, he theorized, the company's solution to a woman member's working with one chemical with known reproductive hazard may well be to "get rid of the woman instead of getting rid of the chemical" (Scott, 1978).

Many see this issue as a problem of sex discrimination, noting, for instance, that when studies linked cadmium to scrotal cancer, or vinyl chloride and lead to chromosomal changes in exposed men, there was no rush to hire women in exposing occupations. In fact, most exclusions seem to have taken place in industries where few women have been employed, and have rarely been suggested in industries where women are a large part of the work force.

As a result of the Pregnancy Discrimination Act of 1978, pregnant women are a protected class under Title VII, and must be treated the same as non-pregnant persons. If a hazard is known to affect the fetus through either parent, an exclusionary policy directed only at women would be unlawful under the policy of the Equal Employment Opportunity Commission. (U.S. Dept of Labor, 1980).

Since reproductive toxins do not confine their action to the reproductive system, and in all cases so far studied have other deleterious chronic effects, the solution to the problem of occupational exposure to toxic chemicals lies in the recognition that exposure entails risk irrespective of gender. Men, women, and their offspring, are all at risk from toxic exposures; we do not know whether these risks are sufficiently different to justify the selection of one of them as needing more protection than the other. In this situation the only responsible and effective solution is to carry out the mandate of the Occupational Health and Safety Act, *i.e.*, to render the workplace safe for *all* who are exposed there—including the fetus.

Realism, however, impels us to admit that the solution will not soon be achieved. In the interim, therefore, those women who wish to become, or who are pregnant, and those men who wish to become fathers, should be given the opportunity to transfer to employment where there is no hazardous exposure, and without loss of seniority or pay. Where union pressure is strong this has, in fact, been the workplace solution for women. (Anon., 1980.) A non-punitive check for reproductive integrity of all workers, with retention of tenure, seniority, and pay rates if a job transfer is indicated, should be a mandatory part of routine workplace health screening tests.

REFERENCES

[1] Amelar, R., L. Dublin and P. Walsh, *Male Infertility*, Philadelphia, W. 1977   B. Saunders.

[2] Anon. Reproductive hazards in the workplace, Chem. & Eng. News, pp. Feb. 11, 28–31. 1980

[3] Barikolata, G. Behavioral teratology: birth defects of the mind. Science 1978   202: 732–733.

[4] Barr, M., C. A. Kelle, W. Rojan, J. Kline. Summary of the workshop 1979   on perinatal and postnatal defects and neurologic abnormalties from chemical exposures. Ann. N.Y. Acad. Sci. 320: 458–472.

[5] Bibbo, M., W. B. Gill and F. Azizi. Follow-up study of male and female 1977   offspring of DES-exposed mothers. Obstet. Gynecol. 49: 1–6.

[6] Cohen, A. N., W. Brown, and D. L. Bruce. A Survey of anaesthetic 1975   health hazards among dentists. J. Am. Dent Assoc. 90: 1291–1296.

[7] Council on Environmental Quality. *Chemical Hazards to Human Repro-* January,   *duction.* 1981

[8] Deknudt, G. H., A. Leonard, and B. Ivanov. Chromosomal aberrations 1973   observed in male workers occupationally exposed to lead. Environ. Physiol. Biochem. 3: 132–138.

[9] Doull, J. D., C. D. Klaassen and M. D. Amdur, eds. *Casarett and Doull's* 1979   *Toxicology: The Basic Science of Poisons.* N.Y. MacMillan Publ. Co.

[10] Gill, W. B., G. F. Schumacher, and M. Bibbo, Pathological semen and 1977   anatomical abnormalities of the genital tract in human male subjects exposed to DES *in utero.* J. Urol. 117: 477–483.

[11] Heinonen, O. P., D. Slone and S. Shapiro. *Birth Defects and Drugs in* 1977   *Pregnancy.* Littleton, Ma. Public Science Group.

[12] Hogarth, P. J. *Biology of Reproduction.* N.Y. John Willey and Sons. 1978

[13] Hunt, V. R., *Work and the Health of Women.* Boca Raton, Fl., CRC 1979   Press, Inc.

[14] Infante, P. F. and J. K. Wagoner. The effects of lead on reproduction. June,   *In* Proc. conf. on women and the work place. Washington, D.C., 1976   SOEH. pp. 232–242.

[15] Mirakhur, R. K., and A. V. Badve. Pregnancy and anesthetic practice 1975   in India. Anaesthesia 30: 18–22.

[16] Odell, W. D. and D. L. Moyer. *Physiology of Reproduction.* St. Louis, 1971 Mo., C. V. Mosby Co.

[17] Scott, R. Reproductive hazards. Job Safety and Health. May 7–13. 1978

[18] Shepard, T. H. *Catalog of Teratogenic Agent.* 3rd ed. Baltimore, Md. 1980 The Johns Hopkins University Press.

[19] Stellman, J. *Women's Work, Women's Health,* N.Y. Pantheon. 1978

[20] Strobino, B. R., J. Kline and Z. Stein. Chemical and physical exposures 1978 of parents: effects on human reproduction and offspring. Early Human Devel. 1: 371–399.

[21] Sullivan, F. M. and S. M. Barlow. Congenital malformation and other 1979 reproduction hazards from environmental chemicals Proc. Roy. Soc. Lond. (B) 205: 91–110.

[22] U.S. DHEW. Guidelines on Pregnancy and Work. DHEW/NIOSH. Publ. Sept., No. 78–118. 1977

[23] U.S. Department of Labor, Equal Employment Opportunity Commis-Feb. 1, sion: Interpretive Guidelines on Employment Discrimination 1980 and Reproductive Hazards. Federal Register, vol. 45, pp. 7514–7517.

[24] U.S. EPA 1977 Air Quality Criteria for Lead. EPA 600/8-77-01. Wash-1977 ington, D.C. 20460, U.S. EPA.

[25] Vander, J., J. H. Sherman and D. S. Luciano. *Human Physiology.* 2nd 1975 ed. N.Y. McGraw Hill.

[26] Wagoner, J. K., P. F. Infante, and D. P. Brown. Genetic effects asso-June, ciated with individual chemicals, in Proc. Conf. (etc.) pp. 100–1976 119.

[27] Whorton, D. M., and T. H. Milby. Recovery of testicular function among 1980 DBCP workers. J. Occup. Med. 22: 177–179.

REPRODUCTIVE HAZARDS IN THE WORKPLACE

A Course Curriculum Guide

written by:
Wendy Chavkin
Ruthann Evanoff
Ilene Winkler
with
Ginny Reath
and the
NYCOSH Reproductive Hazards
Committee

The Reproductive Rights Project of the New York Committee for Occupational Safety and Health evolved from a group which came together in winter 1979 in New York City to take action against employment policies that excluded women on the basis of fertility. This issue had just come to national attention with a report of the four women workers at the American Cyanamid plant in Willow Island, West Virginia, who underwent sterilization in order to keep their recently obtained higher paying production jobs.

Our group was composed of members of the Coalition for Abortion Rights and Against Sterilization Abuse (CARASA), Committee to End Sterilization Abuse (CESA), HealthRight, NYCOSH, and several union activists. Our first activity was to hold a conference in April 1979 which was attended by 250 people. The large amount of enthusiastic participation demonstrated the concern people felt around the issues of exclusionary employment practices and occupational hazards to reproduction.

In response to the need for more information voiced at the conference, we developed this course on reproductive hazards

© 1980, New York Committee for
Occupational Safety and Health

in the workplace. Our goals were both to provide (and demystify) knowledge, and to build organizing skills. We directed our outreach to union activists, but many women's health activists also attended. This reinforced our belief that organizing around this issue can, and must, be done among both trade unionists and members of women's groups and community activists.

The editors of Genes and Gender IV acknowledge their gratitude to the New York Committee for Occupational Health and Safety (NYCOSH) for permitting the inclusion of the following course and curriculum guide. If you would like to use this material as a course, the complete curriculum guide can be obtained from the publisher, The New York Committee for Occupational Health and Safety (NYCOSH) at 32 Union Square East, New York, New York 10003—room 404.

Workers, women and men, are not well protected against the hazards of the workplace. Several reasons contribute to this situation. Management is primarily concerned with profit and this often leads to negligence and cover-ups. Regulatory agencies are often inefficient and lack enforcement powers. Most importantly, many workers are not sufficiently informed to discover the hazardous conditions, and lack the organization to correct them.

The following guide is intended to provide workers with the basic information and organizational know-how they need to take responsible action. The stress is on reproductive hazards for women and men and on the resulting discrimination against women in the work force. Information like this deserves wide distribution and we are pleased to provide yet another channel.

SESSION ONE:
WHAT ARE REPRODUCTIVE HAZARDS

The Female Reproductive System
How it works:
(1) The menstrual cycle is a process governed by a delicately balanced system of hormones. At birth each ovary contains thousands of eggs, of which only about 300 will mature.
(2) Ovulation is the process of releasing a matured egg from

the ovary. Most women ovulate from the same ovary. For many women ovulation goes unnoticed, although some can tell by a twinge or cramp in the middle of the cycle. The egg is released and enters the Fallopian tube. The Fallopian tubes are each lined with small hairlike extensions called cilia which help "carry" the egg down to the uterus.

(3) During each menstrual cycle the uterus begins to develop a thick lining which is prepared in order to protect and nourish a fetus. However, if fertilization does not occur, the lining will slough off along with the egg; this is the menstrual period. The blood and tissue lining passed is usually no more than one-third of a cup, though it seems like more. After the period is over, the hormonal process starts all over again. Approximately 14 days before the first day of menstrual bleeding, ovulation occurs. The menstrual cycle averages 28 days although the length is different from woman to woman. In addition, it may be several years after the first period before the cycle becomes regular. It is important to notice cycle changes (amount of blood, pain) since a change from the individual's normal cycle may indicate a problem.

The Male Reproductive System

How it works:

(1) While at birth girls have all their eggs, men begin to develop sperm at puberty, and continue to generate new sperm throughout their adult life. Sperm are developed in the testes and travel through a long tube, the vas deferens, which exits through the penis. Since the bladder connects with this tube this is also the tube through which the man urinates. Several other organs (seminal vesicles, prostate) connect with the tube and secrete important fluids necessary for the function and mobility of the sperm. When a man has an orgasm, a valve blocks the passage of urine, and he ejaculates sperm along with the other seminal fluids.

Conception and Pregnancy

During sexual intercourse, when a man has an orgasm, 100–500 million sperm are ejaculated. Unless stopped by a barrier method of birth control, a certain amount of sperm pass through the cervix into the uterus and up into the Fallopian tubes. If sperm meet the egg in the tube, fertilization can take place. At fertilization, the sperm and the egg each contribute half of the chromosomes (genetic material). The fertilized egg will then be moved down the Fallopian tube into the uterus, where it attaches itself to the nutrient lining of the uterine wall. This process from fertilization to implantation takes approximately one week. The fertilized egg begins to implant itself into the uterine lining and a series of complex cellular divisions takes place as the embryo develops.

The placenta is a network of tissue and blood, through which the growing fetus derives its nourishment—oxygen. The placenta is connected to the uterus, and bathed in the mother's blood. It is connected to the fetus via the umbilical cord.

How Damage to Human Reproduction Can Occur
A. *Prior to conception*

1. Sexual desire or function can be damaged. We do not know exactly how this occurs, but it can happen to women and men. Exposure to lead reportedly decreases male sexual desire.

2. Interference with the endocrine system. Women and men have different, delicately balanced hormonal systems. If hormone balance is upset in women, ovulation may not take place. For example, women taking anti-psychotic medication do not get periods. If the male hormonal system is disturbed, the production of sperm (spermatogenesis) may be disrupted.

3. Direct damage to the testes, to the sperm producing cells. This may result in the product of insufficient or abnormal sperm. DBCP is a well known example.

4. Chromosomes may be damaged by substances called mutagens, so that sperm or egg will contain abnormal genetic material. This could result in an embryo too defective to live (experienced as a miscarriage) or in a liveborn child with a birth defect. Vinyl chloride is suspected to cause both of these. While both sperm and egg carry genetic material, there is a fairly rapid turnover of sperm, so that if exposure to the hazard stops, new nondefective sperm may again be produced. A woman, however, has all her eggs from birth, so that damage may be cumulative.

Once the genetic material has been altered, this mutation is passed on through the generations. While some mutations may cause obvious changes, others may cause very subtle alterations in the offspring, or changes which may not be apparent until several generations later.

B. *During pregnancy, while the fetus is in the uterus*

1. In general almost all substances in the maternal blood stream (whether ingested, inhaled or absorbed through the skin) pass through the placenta to the fetus. The fetus may be affected very differently than the mother. While occasionally the immaturity of the fetus may serve to protect it (e.g. from a substance that becomes more toxic when metabolized by the liver), by and large, it is more vulnerable. Periods of rapid growth, rapid cell division, organ formation and development tend to show the most devastating responses to harmful agents.

In addition, substances can pass through the placenta to the fetus and cause the child, after a latency period of years, to develop cancer. Such a substance would be considered a transplacental carcinogen. The only known example of this in humans is DES which, after approximately a fifteen year latency period, caused vaginal cancer in young women exposed in utero. There is a strong suspicion that radiation in utero may increase rates of childhood leukemia.

2. Immediately after conception ("embryogenesis"): In the first few weeks the fertilized egg has to begin cell division, finish its journey down the Fallopian tube to the uterus and implant in the uterine wall, where it begins to grow. Any interference with this process is lethal to the fetus, causing early miscarriage (which may be unrecognized by the woman).

3. The first trimester ("organogenesis"): The first twelve weeks of pregnancy comprise the period of organ formation. Hazardous exposure during this period can cause major deformities in the offspring's heart, brain, limbs or other organs. A substance causing this sort of structural defect is known as a teratogen. High doses of radiation in Hiroshima caused those exposed at this stage to be born with small heads and mental retardation. The drug Thalidomide, which caused incomplete development of the limbs, is a well known example.

4. The second and third trimesters: During this period the fetus is developing and growing. Toxic exposure could affect full growth and development. The fetal enzyme systems are

immature and the fetal liver cannot metabolize (detoxify) substances yet.

    5. Anything that causes early delivery or low birth weight places the baby at risk. There is suspicion that lead may cause early delivery. Cigarette smoking causes low birth weight.

C. *After birth*

    1. Toxic substances in the maternal system can be secreted in breast milk. DDT has been found in breast milk samples.

Physiologic Changes of Pregnancy

During pregnancy there are physiologic and physical changes in a woman that affect the way her baby handles toxic substances and her physical needs at work.

(1) Her blood volume increases to one and one-half times its non-pregnant volume. The resulting dilution of red cells to plasma is the "physiologic anemia" of pregnancy and not abnormal. It does make her particularly vulnerable to other factors or agents that induce anemia, e.g. malnutrition (iron deficiency), lead, benzene.

Because of this increase in volume her heart has to work harder, and more blood goes to all her organs.

(2) She also breathes more deeply, achieving higher concentration of inhaled substances in her lungs. Since more blood is going through her lungs, she absorbs more in general into her bloodstream. Thus, her deeper respiration and increased pulmonary perfusion result in her absorbing more toxic substances into her bloodstream. This can harm both her and the fetus.

(3) Greater blood volume and the growing weight of her uterus together increase pressure on her leg veins. This makes prolonged standing or sitting in one position uncomfortable and puts her at greater risk of developing varicose veins in her legs.

(4) The weight of the enlarging uterus puts a strain on her lower back. Again, prolonged standing, or uncomfortable work positions aggravate this.

(5) The increased blood flow to the kidneys causes her to urinate more often. Job restrictions on "bathroom breaks" conflict with this physical need.

Research on Effects of Substances

There has been a lack of research on the specific effects of different substances on reproduction. Much of the research that has been done has been reported in non-English-language journals. Also, the bulk of the research done in this country has been done by industry, or in industry supported studies. While much remains to be discovered, the effects of some exposures are known. Good sources of information on specific hazards are *Work is Dangerous to Your Health* by Jean Stellman and Susan Daum and *Working for Your Life*, by Andrea Hricko.

Substances thought to be hazardous to reproductive function include:

Agent	Sex affected
Anesthetic gases	Women, possibly men
Cadmium	Men
DBCP	Men
Hormones (in pharmaceutical manufacturing)	Both
Lead	Both
Mercury	Women
Pesticides	Possibly men
Radiation	Possibly both
Vinyl Chloride	Possibly both

SESSION TWO: MANAGEMENT'S SOLUTION TO THE HAZARDS

Management's solution to the problem of reproductive hazards is to remove women from the jobs where they are affected.

What's Wrong with This Solution?
It doesn't do anything about the hazards.
It deprives women of jobs.

Do Women Need Their Jobs?
Yes. Women's jobs are no longer only in the home. 46% of American women work outside the home. 45% of married

women work outside the home. 41.5% of women who gave birth in 1972–73 worked during their pregnancies. However, women earn only 57¢ for every dollar men earn. Even at these low wages, 3/4 of the women who work outside the home support themselves or their families, or are supplementing a family income that would otherwise be under $7,000 a year. Women work (while they simultaneously bear and raise children) because they have to.

But Management Claims They're Protecting Fertile Women from Bearing Deformed Children. What's Wrong with That Claim?

Management is assuming that every woman of childbearing age is pregnant, unless proven otherwise, so any substance harmful to the fetus is considered harmful to all women. This assumption is based on the belief that women are not capable of planning their pregnancies or knowing when they're pregnant. It's also based on the sexist assumption that women only work when they are not fulfilling wife and mother roles. Facts prove this wrong. Finally, it assumes that women cannot and should not make decisions about their lives and bodies—the same argument made by the "Right to Life."

Management's real concern is not humanitarian, but, rather, fear of lawsuits brought by affected children. While workers cannot sue their employers for illness or accident (compensation can only come through Workers' Compensation procedures), the malformed child of a worker can sue, if it can be proven that exposure of the parent led to the defect in the child.

Then Companies Must Be Moving All Women Out of Every Job Where There's Danger?

No. Women in traditionally female jobs are not the object of management's concern. Dental technicians, garment workers, beauticians, lab technicians, nurses, operating room nurses and aides are some of those suspected of facing reproductive hazards at their work. Yet no management has said that the fetus is endangered by any of these lower-paying jobs traditionally performed by women.

Who Is Being Excluded?
Women who are moving into traditionally male industrial jobs—jobs with higher pay.

This Kind of "Special" Protection Has Happened Before.
Protective legislation—laws limiting overtime, night work, heavy lifting, etc.—covered women in the U.S. until the late 1960s, when they were ruled discriminatory and repealed. When these laws were passed at the turn of the century, working conditions were very bad. Rather than protecting all workers, women were singled out for "special" protection. The result: women were "protected" out of higher-paying jobs, into lower-paying "women's jobs." Example: telephone operators were exempted from the law so women operators could work nights but higher-paying telephone craft jobs were not. Other occupations exempted from protection included hospital workers and cleaning women (also exempt today from employer concern about reproductive hazards). In other words, this concern was not extended to those lower-paying jobs that traditionally relied on women workers.

What's the Result of "Special" Protection This Time?
Women are kept out of higher paying jobs. Newly established affirmative action ladders are blocked. Women lose their union seniority and men and women workers are made to see each other as competitors, with different interests.

The same thing happens when blacks are denied jobs because they have sickle cell trait, or older workers are denied jobs because they have a higher risk of heart attacks. Any exclusionary policy keeps one group from having jobs and simultaneously keeps another group exposed to hazards.

Does This Policy Leave Men Exposed to Hazards?
Yes. (Refer back to Session One.) It is likely that a substance known to cause damage to the reproductive system or fetus may harm the body in other ways. Lead is an example. Women are being excluded from jobs with lead exposure, but lead also causes nervous disorders, liver and blood diseases. Lead has

also been shown to interfere with sperm production in men. Removing women from a job doesn't remove the hazard. (An example: at the Cyanamid plant where the four women were sterilized, OSHA found that the lead levels actually were dangerous to all workers, male and female.)

Men Can Be Affected by Reproductive Hazards as Well.

Women do not make babies alone, and are not the only ones affected. Management may believe in "immaculate conception," but men play a part in reproduction, and men's systems can also be affected by many of the same substances which affect women.

The Question Has to Be: Why Isn't the Workplace Cleaned Up?

There's no doubt that the American workplace is very unsafe—thousands are hurt and killed in accidents, and thousands more develop occupationally caused diseases. There are many examples of such diseases (black lung, brown lung, cancer) and of such hazards (chemicals, lead, radiation, asbestos). Many of these substances affect the community as well as the workers inside the workplace (Love Canal is an example).

Why Is the Workplace Dangerous?

The hazardous conditions at work result from management's drive for profit. The purpose of production is to make profit. To increase profit, short cuts are taken that make work hazardous. To any boss, profits come first. No matter how well-intentioned the boss may seem, he's interested in profit more than in working conditions. Workers can't look to the boss to protect their interests because their interests and his aren't the same.

Management believes it has the right to control working conditions in order to maximize profit. Attempts to challenge management (by workers and their unions, or by government) always produce conflict. Management fears that open conflict might interfere with production, and therefore tries to divert the workers' attention and manipulate us to agree to the working conditions.

How Does Management Keep Things the Way They Are?

Sometimes they deny that a problem exists, even covering up knowledge of a hazard and hiding the truth. (Examples: Johns Manville—asbestos; Hooker Chemical—Love Canal; U.S. government—Agent Orange).

Another effective way is to divide workers and set them against each other. That's accomplished when management discriminates against one group, so that different groups perceive each other as competitors for jobs. This is the case with reproductive hazards. Management has defined the problem as a "woman's" problem and has tried to turn women and men workers against each other.

What Can We Do to Change Things?

Working together prevents our attention from being diverted and keeps us from being divided. The real conflict is between the boss and the workers, not among different groups of workers. The real issue is that the workplace is hazardous to all of us—those who work there, those who live nearby, and, at times, those who buy the company's products. And it's management's policies that keep it that way.

What's the Strategy?

Organizing at the workplace: developing union health and safety committees and contract protection. Educating ourselves and other workers and getting information about what's happening to us at work. Learning to trust our own perceptions and our abilities to gain the skills to find out what's happening.

Outside the workplace: exerting political pressure to strengthen existing laws regulating workplace conditions and ensure the passage of new ones. Educating the community, and then spreading the word to other concerned and affected groups.

SESSION THREE:
OUR SOLUTION—ORGANIZING ON THE JOB

What is a Health and Safety Committee?

(1) Union committees are created by the union. Official committees are created by the union constitution or bylaws or by membership vote. Unofficial committees can be organized by groups or workers in a particular workplace. In either case, it's important to have the support of the local union, because without it, it's too easy for the employer to get rid of the committee or make it ineffective.

Members of the committee can be elected or appointed, depending on union bylaws. They should be broadly representative, i.e., from different areas of the workplace, different shifts, different groups of workers (in terms of sex, race, age). This enables the committee to know what's going on throughout the workplace and the workforce and will increase workers' trust in the committee if they feel represented.

(2) Joint union-management committees have representatives from both workers and management, and are no substitute for local union committees. Even under the best contract, the powers of a joint committee are limited. No committee that is restrained in this way can do its job properly. However, these committees can be forums for expressing the union's position to management. The union should participate where they exist. The union representatives should be selected by the union, never by management, and should have a clear idea of the union's position on the issue being discussed. Joint committees should have equal representation of both sides; rotating or co-chairs (between labor and management), jointly developed agendas and records of inspections and meetings signed by both parties.

What Does a Health and Safety Committee Do?

(1) Protect the workers. Make sure existing work rules and safety standards are maintained, and that contract protections on health and safety are carried out. Handle health and safety grievances, or train and assist stewards to handle them. If possible, remove workers from unsafe jobs and shut them down.

(2) Inspect the work area to discover problems and hazards.

To do this the committee must include people who are familiar with all the jobs. It must have continuing contact with the workforce on a day to day basis in order to learn quickly about problems as they develop.

(3) Collect information: on accidents, hazards, work practices, and workers' health problems. To do this the committee must keep records on accidents and look at them for patterns or clues to problems. The committee must have access to health records, information on substances used, OSHA records and monitoring data, and examine all this to detect patterns in the workplace and in the workforce.

(4) Educate and inform the workers. Arrange health and safety training for workers, union stewards and officials, as well as for committee members. Hold open meetings to discuss problems and keep people informed about what's happening. Publicize the committee's work. Respond to people's complaints.

(5) Correct hazards if it has the power to do this directly. Look for solutions by working with union staff, COSH groups, medical consultants, etc.

(6) Organize the workers to develop and fight for solutions to the problems. A strong committee may already have won the contractual right to correct problems directly. More often, the committee will need to mobilize the workers to collect the information, develop solutions, and demand that management take corrective action.

(7) Involve OSHA or other government agencies, if necessary, by filing complaints and following up on OSHA inspections or other procedures.

How Can You Set Up a Health and Safety Committee?

1. Convince the union that it's needed. Unions are increasingly seeing the importance of health and safety, but in the past many unions have been reluctant to deal with these problems. In bargaining, health and safety often still takes a back seat to money demands. (Discuss why this is so.) If your union isn't too interested you can: (a) Educate members and officials about the problems faced by your particular workforce. (b) Organize rank and filers and stewards to publicize the problems and initiate grievances. (c) Involve workers from different areas of the workplace so it doesn't look like an issue that just af-

fects a few people. (d) Set up an informal committee and try to get union backing. (e) Get help from COSH groups, labor schools, etc.

(2) Get it in the contract. The best way to have a committee recognized by management is to negotiate it in the contract. Employers are legally required to bargain over health and safety issues, so even if the contract does not have language on health and safety, stewards can still grieve violations. The contract can also define committee rights to include: (a) Power to correct hazards on the shop floor; (b) Mobility (the right of the committee members to leave their work on company business; the right of the committee to full access of the workplace); (c) Pay for committee work on management time; (d) Access to company records.

(3) If you don't have a union: Your power is much more limited and you will need to rely much more on government agencies (next class). This issue could encourage people to organize a union.

What Provisions Besides the Establishment of a Health and Safety Committee Can Be Included in a Contract?

(1) General duty clause which spells out management's responsibility to provide a safe, healthful workplace. Like a union recognition clause, this provides a contractual basis for grievances on health and safety issues.

(2) Management responsibility. The threat of being sued has scared away some unions from taking on any responsibility for working conditions. It's important for the contract to specify that it is the responsibility of management to provide safe, healthful working conditions.

(3) Right to know. A "right to know" clause ensures access to the workplace and to information. It includes access to labels of substances, so that workers know what they're working with; committee or union access to records on accidents and illness; may include access to the entire workplace by the union health and safety committee; periodic health evaluations by outside hygienists.

(4) Specific provisions including work rules, agreements on monitoring, use of personal protective equipment, installation

of controls, safety procedures, etc. This could also include a provision for handling health and safety disputes, how they fit into grievance procedure. Since arbitrations on health and safety issues are difficult to win, these disputes should be kept away from procedures leading to arbitration, without sacrificing the right to grieve.

(5) Right to refuse/right to shut down a job. This is the hardest to win but it gives weight to all the rest. You are now protected by the courts if you refuse a job that involves imminent danger to life. However, refusing a job often leads to another worker being brought in, and usually leads to unfavorable arbitrations. The right to shut down ends the whole job and protects all the workers.

(6) Protection for workers. Workers with occupational diseases or injuries, or workers who are at risk for particular reasons (pregnant women, for example) must be protected. Clauses can include: light duty; the right to reassignment to another job with no loss in pay, benefits or seniority; long-term disability protection; and company-paid medical assessments and treatment.

How Can the Boss Try to Turn the Work of the Committee Around?

(1) Petty harassment. For example, the boss says radios keep people from hearing warning bells and the company forbids them. Workers direct anger at the union. Avoiding this needs shop floor creativity.

(2) Unpopular safety measures. Rather than using engineering controls, the company will issue personal protective equipment and discipline people for not using it. (Note: OSHA says personal equipment can be used only in emergencies, if there is no other way to protect workers, or while changes are being made.)

(3) Retaliation against workers. For example, the boss harasses active committee members. You can retaliate on the shop floor, or you can complain to the NLRB, OSHA or EEOC. This is protected activity.

(4) Exclusionary policies. The company excludes a group of workers, supposedly to protect them. A group might consist

of fertile women, black workers, older workers (who can no longer pass medical exams because of damage the job has already done to them). This makes workers suspicious of medical exams, which is a problem since exams are important. This is a reason why the union must have equal access to tests and records, and a reason why the union may want to set up its own medical screening programs (as well as to guarantee confidentiality). All exclusionary policies separate and divide workers and threaten unions by destroying solidarity and breaking down the seniority system, permitting arbitrary hiring and firing.

(5) Threatening to close the plant. The boss claims cleaning the plant up is too expensive, and threatens to move. This is a tough one. You have to assess if it's a scare tactic, and if so convince people that's all it is, or if he's serious. If he is, you have to understand health and safety problems aren't the real reason and organize your fight accordingly.

A Brief Guide to Organizing Strategies
1. Define the problem
2. Define the solution
3. Assess what can be done:
    a. Can you get information on the problem?
    b. Is the problem widely or deeply felt? (This will determine whether people will be willing to act.)
    c. What are the committee's strengths and weaknesses?
    d. What will the boss's response be?
    e. Can a victory be won? (Even a partial victory builds confidence and credibility for the committee.)
4. Set a timetable and deadlines
5. Decide on strategy and tactics
6. Organize the shop
    a. Get support of people in the shop
    b. Get their input and get them involved
    c. Confront apathy. Understand what it is and how it affects people.
    d. Get commitments, get to know people and use people's skills. Be sure everyone shares in different levels of work.
7. Publicize your victories.

Sample Problems for Workshops
I.   Shop/department—Hospital X-ray therapy.

Workers—total 24 (widely varied ethnic and racial background) 3 male MDs; 2 women RNs (not in union); 4 women LPNs; 12 technicians (8 women, 4 men—at greatest risk of exposure); 3 women clericals (all in same union).

Background—There have been 2 miscarriages and 1 mastectomy among workers over the past few years. There is no health and safety committee.

Problem—Workers wear film badges to measure x-ray exposure which are turned in for monthly readings. An RN raises the issue that workers are never given the reports of exposure readings.

Questions—Should they form a committee? If so, who should be on it? What should they do with the exposure reports? What else needs to be done?

II.   Shop/department—2 departments in a 10-department unionized chemical plant. Department A mixes chemicals used to make tranquilizers on the day shift only. Department B manufacturers tranquilizer capsules on two shifts.

Workers—Department A—15: 13 men and 2 women (all higher paid than Department B), includes 1 supervisor. Department B—60: 45 women and 15 men, includes 5 supervisors.

Background—There is a six month old health and safety committee with one representative from each department. It is chaired by management which has six representatives on the committee.

Problem—Management announces that women in Department A must bid for other jobs or be sterilized because of possible risk to a fetus—no seniority or rate retention guarantees.

Questions—What should the workers do? Consider expansion of health and safety committee or formation of subcommittee; the role of "Right to Know" legislation; affirmative action, etc.

The rest of the problems are ideas; they can be expanded as above.

III. Two wives of male operating room workers in your hospital have recently had miscarriages and a couple of women report difficulty becoming pregnant. You suspect this may be caused by conditions in the operating room. What do you do?

IV. All the women working in a pharmaceutical plant have been excluded from working with hormones because the company says it will endanger their reproductive abilities. These jobs pay 55¢ an hour more than any other job in the plant. What do you do?

V. Some of the clerical workers in an animal laboratory discover they are having similar menstrual problems. You wonder if it might be job related but don't have a union. What do you do?

VI. You work in a nuclear power plant and are worried about radiation endangering yourself and your family. A local anti-nuke group is demanding the plant shut down, and many people you work with are afraid they will lose their jobs. What do you do?

SESSION FOUR:
HOW TO LOOK AT YOUR WORKPLACE

Why do You Have to Look at Your Workplace?
You have to identify problems before you can develop strategies to solve them.

What Hazards Can You Spot Right Away?
Spills, fumes, dusts, loud noise, faulty machinery.

What Do You Take for Granted That Might Also Be Hazardous?
Lighting, air conditioning sounds, stress, broken furniture, wires on the floor.

Are These the Only Hazards?
No. These hazards which are easy to spot may not be the only ones, or the most serious. Small continuing insults may cause chronic illness. The little things that you're used to and take for granted are the ones for which you have to be particularly alert. That's why you have to look at your workplace with new eyes.

A Systematic Way to Look at Your Workplace.
This is the first step towards getting yourselves organized to deal with health and safety. You should make a floor plan of your workplace, a flow chart of work and have a description of the work process. A checklist will help as you look and also helps with record keeping. These records are important for yourself, and will also help an inspector if you get an OSHA inspection.

Talk to People.
The workers are the experts. A lot is happening at any workplace. Each worker knows her/his area best—the people you work with can tell you a lot about what's happening at their jobs. So looking at the workplace means talking to the people who work there and going through the work process step by step.

Use Your Senses.
Your eyes can see dust, powders, safety hazards.
Your ears hear noise.
Your nose smells odors.
Your skin feels heat or cold.
All these can be clues to exposures.

But Senses Have Limitations.
You can get used to a smell, or a dangerous substance may not have an odor. You can adjust to noise or your hearing may be dulled. Many things are too small to see (such as dust) or

the lighting may make it hard to see. Use your senses, but don't rely on them alone.

Examples of Work Exposures and Conditions That Might Be Hazardous:

Offices—photocopier, lighting, video display terminals
Hospitals—heat, heavy lifting, infection, stress
Factories—sanding, welding, painting, open vats, slow leaks

The way substances are brought in, stored and sent out is also important. Talk to the people doing the job to find out what's involved in the process.

Check Out the Workforce Too.

You have to look at the people working in the workplace. You're looking for: transfers, absenteeism, injuries, accidents, illness, vague symptoms. If there have been accidents, what caused them? Is there any pattern that might indicate a health problem?

Checklist for workers' problems will include questions about:

(1) Does the person have any known disease, e.g. high blood pressure, diabetes, bronchitis, etc.?

(2) Does the person have other health complaints, e.g. headaches, rash, cough, backache, fatigue, urinary problems?

(3) For women: Has she had difficulty conceiving a child?
Has she had changes in her menstrual cycle?
Has she had miscarriages?
Has she had problems with pregnancy?

For men: Has he had trouble conceiving?
Has he ever been told of trouble with his prostate?
Has his wife/partner ever had any of the above problems?

Do any of their kids have any health problems, e.g. leukemia or any type of cancer? learning problems/retardation? diagnosed birth defects?

This is a very general and preliminary list, but it will help uncover general patterns. There are many resource people who can help you with more detailed health questionnaires for more in-depth investigation.

A new provision by OSHA states that workers must be allowed to see the records the company keeps. You can now follow up on the health of the people in your workplace through medical records. Every employer covered by OSHA also has to post the 200 log that records all accidents. It's a violation if the log is not posted or is incorrect.

In Addition to Looking for Hazards and Health Problems, Look at What Has Already Been Done to Control Hazards.

Methods to control hazards are generally divided into three categories:

> engineering controls which change the process
> administrative controls which change the way the worker does the job
> personal protective equipment which attempts to prevent the process/substance from doing harm.

Engineering controls are the best because they fix the workplace and eliminate the hazard. The technology exists to make most of the changes but management often won't do this (refer back to Session Two for reasons).

Administrative controls include rules about how long a worker can be exposed, or under what conditions (e.g. rotating shifts, time on and off a particular job). These do more to fix the worker than actually fix the workplace. They just measure exposure received but don't stop it.

Personal protective equipment (PPE)—all the respirators, earplugs, suits, etc.—are all ways to fix the workers without fixing the workplace. PPE is not the answer. Most are uncomfortable and some kinds can cause their own safety and health problems. OSHA says PPE can be used temporarily while there is a plan in effect to engineer the problems out permanently.

Check the Steps Taken for Control.

(1) Are the safety procedures used? appropriate? maintained properly?

(2) Does the PPE fit? Is it appropriate? maintained properly? Is there a plan to engineer the problems out?

(3) Are there showers? changing rooms?

(4) How is the housekeeping? maintenance?

Once again, the worker is the expert. Talk to people—the person doing the job knows how well the safety procedures work, how often the area is cleaned, how comfortable the PPE is.

What Can You Learn by Looking at Your Workplace?

Patterns of health problems and hazardous conditions may emerge. You may also identify questions about specific substances or conditions which need further investigation.

SESSION FIVE:
OSHA AND THE EEOC

I. Occupational Safety and Health Administration (OSHA)

What Is OSHA?

The Occupational Safety and Health Act was passed in 1970 in response to the Black Lung movement and pressure from unions. Its purpose is to ensure that workers have a safe workplace by setting standards, conducting inspections, correcting hazards. Under OSHA, every employer has a "general duty" to provide a workplace free from occupational hazards—this section of OSHA is called the general duty clause.

The Occupational Safety and Health Administration (part of the Department of Labor) is responsible for setting and enforcing standards, performing inspections and monitoring compliance. The National Institute for Occupational Safety and Health (part of the Department of Health and Human Services) does the research on which standards are based.

Political interference, along with understaffing and underfunding, has prevented OSHA from being highly effective.

How Does OSHA Get into a Workplace?

The grounds are complaint, accident, imminent danger, random inspection, re-inspection. Originally OSHA inspectors could come in without advance notice. Unfortunately, a recent court decision now permits employers to refuse entry unless the inspector has a court order. This gives the employer time to clean up hazards before the inspection.

How Do You File a Complaint with OSHA?

Regular complaint forms are provided by the agency. These must be filled out explicitly—number of people affected, where, when. This helps OSHA evaluate the complaint and the inspector will come better prepared. The complaint can be completely confidential, and you should always request confidentiality. Section Eleven-C of the OSHA Act protects you from harassment by your boss for filing a complaint. If your complaint is accepted by phone, follow up in writing.

What Happens in an OSHA Inspection?

(1) What is the purpose of the inspection? According to the law, it is to evaluate workplace conditions by comparison with standards set for that particular type of workplace.

(2) When is the inspection? It should take place when the problem occurs, if necessary on a particular shift.

(3) What is the opening conference? This is a meeting of the employer, union representative and inspector. It begins the inspection and should last no longer than an hour, so that the employer can't use the time to clean up the hazards. The union representative is paid at the regular rate for time spent in the conference and inspection. If the union representative doesn't attend the opening conference, or if meetings with employer and union representative are held separately, the OSHA inspector will supply a summary of the meeting.

The union representative should keep her/his own records of this meeting, and of the entire inspection. In the conference the problems should be presented, along with grievance records and written documentation of other problems. This can help prove lack of good faith by the employer, and increase the penalty.

(4) Who goes on the inspection? The inspector and the union representative. If there is more than one union, each covers their own department. If there is no union, one worker can be chosen and/or the inspector can talk with any or all work-

ers privately. Workers have a right to talk privately to the inspector on a confidential basis.

(5) What should the union representative do during the inspection? Familiarize the inspector with the workplace, point out problems, make sure that the inspector talks with people of both sexes from different departments. Be sure the inspector sees normal working conditions and looks at the 200 log (the record of accidents) which must be publicly posted. Record how measurements are made—what kind of instruments did the inspector use, samples were taken over what time period, what were the results?

(6) What happens at the end of the inspection? There is a closing conference to review the inspection and specify hazards. The union representative should repeat the history of the problems, let the inspector know your ideas about solutions, press for an early correction date and describe discrimination problems. If you suspect that the inspector is not going to cite the employer for the violation, ask why. As with the opening conference, if the inspector meets separately with the employer and union representative, the inspector will provide written summary of each meeting.

What Happens After the Inspection?

(1) Evaluation by the OSHA office. If the employer is cited, the citation must be posted. It will describe which standard is violated, the abatement date (date by which hazard must be corrected) and the penalty (if any).

(2) If the employer contests the citation, hearings will be held. The union should file for "party status" by writing a letter to OSHA after the inspection, requesting a copy of the citation, and requesting party status if the citation is contested. This allows the union access to records and participation in all conferences.

Does This Procedure Work?

People get discouraged because OSHA moves slowly. The committee needs to follow up, keep track of dates, letters, etc., and keep people informed of progress.

OSHA's power of enforcement must be strengthened. At the same time, we need to rely on union strength, and not depend on OSHA.

II. Equal Employment Opportunity Commission (EEOC)

What Is the Equal Employment Opportunity Commission?

It is the administrative agency which enforces the provisions of Title VII. It publishes guidelines defining what constitutes discrimination under Title VII, and handles complaints of discrimination. Title VII of the 1964 Civil Rights Act is a federal law prohibiting discrimination by private employers on the basis of race, national origin or sex.

Why is Title VII important? Until/unless the ERA is passed, this is the main basis for federal court actions on most aspects of sex discrimination in employment.

How Do You File a Complaint with the EEOC?

Fill out an official complaint form. Name everyone in the original complaint or you can't sue them later. Make the claim as broad as possible and name everything and everyone that's discriminatory. don't trust the EEOC counselors. They have reduced the previous huge backlog of cases by discouraging people from filing claims.

What Does the EEOC Do with Your Complaint?

They have 90 days to act. They can either (a) investigate and dismiss or (b) investigate and find cause for your complaints. They then will attempt to conciliate. If they won't or can't settle the case you have a right to sue in federal court (the EEOC will send you a "right to sue" letter) but you need a lawyer to do that.

Local agencies may have policies that interact with EEOC—you should check them out.

Is Discrimination against Pregnant Workers a Violation of Title VII?

Yes, since April 1979. Federal law now prohibits (a) discrimination in hiring, firing or promotion because of pregnancy; (b) discrimination in disability benefits for pregnancy or childbirth; (c) discrimination in medical payments for childbirth or pregnancy or in medical insurance coverage. However, the law permits private employers to exempt abortion coverage from medical benefits.

Why was this law passed? In response to pressure from women's groups and unions after the Supreme Court ruled in 1976 that discrimination against pregnant women was not sex discrimination.

Is the Exclusion of Fertile Women from Certain Jobs a Violation of Title VII?

No court has yet ruled on this, but the EEOC has issued guidelines on reproductive hazards and the exclusion of women. The guidelines recommend that women not be excluded unless reputable scientific evidence proves that the substance involved affects fetuses only through women. Once that's proven (very unlikely, see Session Two), the employer can exclude only pregnant women, not all fertile women. Employers also must show that there are not feasible alternatives to reduce exposure to hazards, and that they are in compliance with OSHA guidelines and the OSHA general duty clause.

What are the problems with the guidelines? They permit discrimination (exclusion of pregnant women) until scientific data is collected and allow the employer two years in which to collect such data.

In theory, the EEOC guidelines and OSHA together mandate a safe and healthful workplace for all workers without discrimination on the basis of race or sex.

Is Going to Court a Good Strategy?

Court victories can be important. There are, however, drawbacks to the legal approach. A court battle is hard to win, takes a long time, forces you to depend on experts (lawyers), and may drain energy and money from other areas of struggle. A legal victory may be very helpful but it cannot be a substitute for the primary task of building worker and union strength.

BIBLIOGRAPHY

Journals

Feminist Studies, v. 5, n. 2, Summer 1979. Special issue on reproductive hazards and the politics of protection. (Single copies $4.00 prepaid from Managing Editor, Feminist Studies, c/o Women's Studies Program, University of Maryland, College Park, MD 20742.)

Preventive Medicine, v. 7, n. 3, Sept. 1978. Special issue: "Forum: Women's Occupational Health: Medical, Social and Legal Implications. (Available at cost from Academic Press, 111 Fifth Ave., New York, NY 10003, or in medical libraries.)

Books

Chavkin, Wendy. *Occupational Hazards to Reproduction: An annotated bibliography.* (Available from Department of Social Medicine, Montefiore Hospital, 111 E. 210th St., Bronx, NY.)

Hricko, Andrea. *Working for Your Life.* (Available from Labor Occupational Health Program, Center for Labor Education, University of California, 2521 Channing Way, Berkeley, CA 94720. $6.00)

Hunt, Vilma R. *The Health of Women at Work. Occasional Papers No. 2.* (Available from Program on Women, Northeastern University, 619 Emerson St., Evanston, IL 60201. $6.00)

Stellman, Jeanne. *Women's Work, Women's Health: Myths and Realities.* (NY: Pantheon, 1977. $3.95 paper)

Stellman, Jeanne and Susan Daum, MD. *Work is Dangerous to Your Health: A Handbook of Health Hazards in the Workplace and What You Can Do About Them.* (NY: Vintage, 1973. $3.95)

Films

Song of the Canary. (Available from New Day Films, PO Box 315, Franklin Lakes, NJ 07417. 16 mm, color, 58 min. Rental $65.00 + $5.00 handling)

Working for Your Life. (Available from Labor Occupational Health Program Films, 2521 Channing Way, Berkeley, CA 94720. $65.00 rental.)

COSH Groups

CACOSH, 542 South Dearborn, Rm. 502, Chicago, IL 60605. (Chicago area)

ECOSH, 655 Castro, No. 6, Mountain View, CA 94041. (electronics manufacturing)

MaryCOSH, PO Box 3825, Baltimore, MD 21217 (Maryland)

MASSCOSH, 120 Boylston St., Rm. 206, Boston, MA 02116
276 High St., Holyoke, MA 01040 (Massachusetts)

NJCOSH, 701 E. Elizabeth Ave., Linden, NJ 07036 (New Jersey)

NYCOSH, 32 Union Square E., Rm 404, New York City 10003 (New York City area)

NCOSH, Box 2514, Durham, NC 27705 (North Carolina)

PHILAPOSH, 1321 Arch St., Rm 201, Philadelphia, PA 19107 (Pennsylvania, Delaware)

RICOSH, 371 Broadway, Providence, RI 02909 (Rhode Island)

TENNCOSH, c/o Center for Health Services, Station 17, Vanderbilt Medical Center, Nashville, TN 37235 (Tennessee)

WNYCOSH, 59 Niagra Square Station, Buffalo, NY 14201 (western New York)

WISCOSH, 805 South 5th St., Rm. 201, Milwaukee, WI 53204 (Wisconsin)

(Contact NJCOSH for new additions to the COSH network.)

SUGGESTED READING LIST

Session One
*Hricko. *Working for Your Life*. Chapters 6, 7, 3. Chapters 9–18 on specific hazards.

*Hricko. *Working for Your Life*. Chapters 1–5.

Feminist Studies. pp. 233–246, 274–285, 302–307, 310–325.

Preventive Medicine. pp. 394–406.

McGhee, Dorothy. "Workplace Hazards: No Women Need Apply," *The Progressive*. 20 October 1977, pp. 20–25.

Scott, Rachel. "Reproductive Hazards," *Job Safety and Health*. May 1978, pp. 7–13.

Lehman, P. "Women Workers and Job Health Hazards," *Job Safety and Health*. April 1975.

Edminston, Susan and Julie Szekely. "What we must know about health hazards in the workplace," *Redbook*. March 1980, pp. 33–

Session Three
How To Look At Your Workplace. Available from Occupational Health and Safety Group, Urban Planning Aid, 639 Massachusetts Ave., Cambridge, MA 02139, $1.00.

IAM Guide for Safety and Health Committees. Available from the International Association of Machine and Aerospace Workers, 1300 Connecticut Ave., NW, Washington, DC 20036.

Session Four
Sample checklists: *Hricko. *Working for Your Life*. pp. E18–21, E22–23, E24–29.

How to Look At Your Workplace.

A Workers Guide to Documenting Safety and Health Problems. Available from Labor Occupational Health Program, 2521 Channing Way, Berkeley, CA 94720.

Session Five
OSHA: Your Workplace Rights in Action: #3029—Job Safety and Health: OSHA Inspections are only the beginning; #3021—Workers' rights under

OSHA; #3032—You have a right to protect your life on the job. That right is called Eleven-C. Pamphlets available from OSHA Regional Offices.

How To Use OSHA: A Workers Action Guide. Available from Urban Planning Aid, address above.

A Workers' Guide to NIOSH. Available in Federal government bookstores.

*Preventive Medicine. pp. 372–384.

*Feminist Studies, pp. 247–273, 286–301.

*See bibliography for complete information on all readings marked with an asterisk (*).

The following articles examine the effects of discrimination on women, as they attempt to cope with the extant social structures.

Gold's paper describes a more overt form of the socialization of professionals into a view of women, which is operant in the training and continuing education of medical doctors.

Bram investigates how professional psychological literature assumes a point of view and thereby socializes its students and practitioners to accept certain premises. Child-bearing is the case in point.

Fooden discusses the mental health problems generally ascribed to women and investigates the epidemiological data which underlie the assumptions, as well as suggested treatment methods.

Vroman's paper deals with a select group in double jeopardy because of age and sex, the older woman. She discusses how discriminatory practices mitigate against opportunity for help and perpetuates the status quo.

The paper by June Christmas is concerned, not only with health delivery to women, but of the effects of discriminatory policies on the health prognosis of minorities and poor.

SEXISM IN GYNECOLOGIC PRACTICES

Marji Gold, M.D.

This article will describe some of the sexist attitudes that are prevalent in medical practice and how these ideas are taught to new doctors and reinforced in experienced practitioners. Some examples of how sexism and racism lead to well-documented inferior care for women and especially minority women will be presented. Lastly I will try to suggest some approaches to changing this racist and sexist system.

The perpetuation of sexist ideology is not just the province of "sexist doctors." Rather, administrative and governmental policy mandates and reinforces sexist attitudes and practices (InCAR, 1979). The current cutbacks in medical care are a good example of this. In a society based on the profit system, non-profitable expenses such as services to minorities and women, take a back seat to investment in military production and support of private industry. Each physician is a product of the system which trained and supports him (or her), and attacks only on physicians are not enough; the larger society must be criticised and changed.

This paper is based on my personal experience and that of my patients, some lay books and articles as well as the medical literature.

I am a family physician, not a gynecologist; however, I work closely with gynecologists. I have spent my ten years of clinical experience—as a student, a resident, and now an attending—working in hospitals in New York City. I have worked at both the public city hospitals and the affiliated voluntary (private "not-for-profit") hospitals, so I have a broad perspective on what happens to women of all social strata, and how minority women receive the worst care of all.

Some literature written for the general public that I have utilized, most notably, *The Hidden Malpractice* by Gena Corea; *No More Menstrual Cramps and Other Good News* by Penny Budoff; and *The Public Health of Racism in The United*

States by the International Committee Against Racism have extensive data and analyses of the sexism and racism in medical practice.

The documentation in the medical literature was more difficult to find but was present in formal reports from reputable medical researchers and epidemiologists. These data clearly supported the conclusions I had come to from my personal experience both as a doctor and a political activist which I have been during most of my medical career.

In many ways I believe that having a political perspective and ties to an organization has been the mainstay of my surviving in the medical system. Anti-racist and feminist ideas do not predominate in medicine, and I find it useful to see medicine as a reflection of society, in order to understand where typical medical attitudes originate and who supports them.

Before proceeding further, I would like to define some terms. *Gynecologic care* refers specifically to medical care of the reproductive organs of women and performed by trained specialists, i.e., gynecologists. However, because many women don't know this and because women focus in on reproduction and sex as they are taught in our society, they use a gynecologist for their general medical care. Because of this special relationship between women and their gynecologists, and because *all* doctors learn their attitudes towards women from gynecology professors, I have expanded the definition of gynecologic care to include all general medical care of women.

I deal with racism and sexism as the daily reality of second class treatment for minorities and for women, dependent on but not limited to attitudes: i.e., statistically inferior jobs, lower pay, worse medical care and so on. These sexist and racist attitudes have been defended by sociobiologists, who justify the inferior status of women and minorities in our society by attributing socially determined behaviour to genetic differences. Although many scientists have discredited the sociobiologists' notion of genetically determined human behaviour, the attitudes that these "theories" support continue to prevail. Thus natural allies are kept from uniting to work for improvement in the workplace and in medical care.

I also want to make clear that racism and sexism are not restricted to medical treatment of women as patients. Women health workers, and especially the non-professional minority women who make up the largest segment of workers in the health care "business" (Corea, 1977), must deal not only with what happens to their patients but also what happens to themselves because of anti-minority, anti-women ideas and practices. This paper however will consider only those issues that affect the treatment of women as patients.

The main issue in recent ideology is that women are essentially stupid or that there is something in their genes that makes them less intelligent than men. In addition, they are "hysterical," a medical term that is roughly equivalent to emotional but that also connotes and reinforces an absence of intelligence. Women can't think or understand, they just *feel*. One significant generalization made from the acceptance of the above sexist ideas is that women cannot give a good medical history, i.e., the doctor can't trust or believe anything she says about her condition or illness (Howell, 1974).

Another assumption about women is that they are only really interested in getting pregnant and having babies. Any woman who denies these desires is considered dishonest or "unfeminine." Extrapolation from this concept tells us that a woman has painful menses because each period signifies that she is not pregnant and/or she is dissatisfied with her feminine role. While it is true that the ability to become pregnant is a function of women's anatomy and physiology, an interest in having children is surely not limited to women. In addition, it is assumed that post-menopausal women are uniformly depressed and cannot function effectively on the job. This depression is believed primarily to come from the absence of children in the home and an inability to conceive more. The ultimate conclusion of the medical profession is that women are crazy and unreliable; unlike men, women don't have *real* illnesses.

How do these ideas get disseminated and integrated into a physician's style of practice? First of all, because all medical students have gone through 16 years of education in a sexist system, they already believe some of the premises; medical education just puts these ideas in a medical context. Written materials—most notoriously, obstetrics and gynecology textbooks but also medical journals—are replete with sexist statements. The following quote from an obstetrics and gynecology textbook (Wilson and Carrington, 1979, p. 5) is a description of what a physician should look for when a patient comes into his office:

"Character traits are expressed in her walk, her dress, her make-up, her responses to questions and in almost every action . . . the observant physician can quickly make a judgement as to whether she is overcompliant, over demanding, aggressive, passive, erotic or infantile."

Could she have any good characteristics? The same text (Wilson and Carrington, 1979, p. 105) in discussing menstrual cramps makes the following statement:

"Whatever may be the cause of painful uterine contractions, a strong psychosomatic influence plays an important role in primary dysmenorrhea. There is usually a marked hypersensitivity to pain and in many cases there are evidences of personality disorders. . . . Dysmenorrhea . . . often occurs in an individual who has been psychologically conditioned to this disorder." Another paragraph continues:

"Most females with premenstrual tension are depressed (10–20 percent of women in the fourth decade). The reason for this depression may be anger over their discomfort and the guilt they feel because of their behaviour within the family. The husband often can be helpful by not being too sympathetic and being firm in an understanding way."

In other words, a woman who suffers from menstrual cramps is one who is emotionally disturbed and the way to handle the situation is to treat her like a naughty child. Many similar examples can be found in other texts and in medical journals.

Advertisements of drug companies in the medical journals are another way in which sexist ideas are perpetrated. The following are three advertisements for essentially the same drug marketed under different brand names.

(1) One showed a man exercising, bending and lifting and the ad suggested that this drug be used as a muscle relaxant.

(2) A second ad pictured two women shopping and smiling as a result of taking the drug for "non-psychotic anxious depression." The drug company further proclaims that "broad symptomatic relief is achieved of symptoms including agitation, anxiety, anorexia and feelings of guilt and worthlessness." Of course these are only symptoms women have and the drug will allow them to be *free to go shopping*.

(3) The third pictures a smiling woman at a picnic reading to her child and the ad is headlined "Freedom To Think Clearly." This drug is also advertised as an anti-anxiety medication that will allow a woman to think clearly so as to be able to perform her feminine role of reading to her child.

The obvious conclusion to the practitioner is that while men need drugs because of symptoms having a physiological basis and caused by muscular activity, women's symptoms all have an emotional basis and are not physiological.

Another important way in which sexist and racist attitudes are developed in medical students and residents is by observing their role models, the attending physician. The following describe extreme examples because the more subtle ones are harder to relate. When attendings do rounds with residents and students they are teaching a way to approach the patient. An extreme example of sexism is the attending gynecologist who taught students to rub the woman's clitoris before doing a speculum examination in order to have her lubricate and thus have an "easier exam." Other examples include not respecting a patient's privacy, i.e., bringing in six people to watch a pelvic exam and even repeating it; talking about a patient to the residents and students as if she was not there; and not explaining procedures and side effects to patients. Another prevalent attitude based on the notion that women are stupid and emotional, is that the doctor "takes care" of her and treats her like a young child. Doctors become offended when women want to participate in their own health care, for example when a woman in labor asks what medication she is receiving and why.

The result of these sexist ideas and practices is poor medical care, on both an individual and an institutional level. The following are only a few examples documented from the literature. Note that minority women are affected even more.

(1) 7700 American women die per year of cervical cancer. Death from this disease is preventable if caught early and detection is possible through a PAP smear (Fruchter, Boyce and Hunt, 1980).

(2) Hysterectomies (surgical removal of the uterus) are the second most common surgery performed in the U.S. Many "second opinion" and retrospective studies found one-third of these unnecessary (McCarthy and Finkel, 1980). The attitude

of many physicians is that the uterus is not needed after a woman's reproductive period is over, and therefore why not remove it for even minor problems that could be treated in other ways. This sometimes extends to removal of the ovaries of women over 50 "since you will be going into menopause soon anyway."

(3) IUD's are a common contraceptive, widely prescribed. Recent studies have shown that their use is often associated with infections that then lead to sterility (Buchman, 1981). In addition they are 90–94 percent effective compared with 99 percent effectiveness of pill or diaphragms. They are particularly prescribed for minority women because of racist and sexist rationalizations of the unreliability of minority women in using pills and diaphragms.

(4) Sterilization abuse is not uncommon with doctors using threats to force women to agree to tubal ligation. "I won't care for you if you refuse to sign this consent form" is one way in which this is accomplished.

(5) Abortion funding has been cut which affects poor women in general and minority women in particular. In addition the campaign to illegalize abortions has been considerably stepped up (Gold and Cates, 1979).

(6) Funds to help pregnant women receive proper nutrition and prenatal care have been drastically reduced. Drugs are often prescribed to pregnant women without sufficient regard to the effects on the fetus (InCAR, 1978).

(7) Mood-altering drugs are used by 50 percent of the women in the U.S. Women are 70 percent of the habitual users of minor tranquilizers and 72 percent of the users of antidepressant drugs (Chambers, 1971). This is again the result of the physician's approach that men have real illnesses while women's complaints are because they are "crazy." Women either suffer from "premenstrual tension" or "menopausal depression" and these are all "in the mind" and therefore tranquilizers are the way to treat them.

(8) The city supported hospitals which service a high risk population do not have sufficient numbers of attending obstetricians and gynecologists—in addition because of a lack of facilities and supporting staff, tests that are necessary for proper diagnosis and treatment are not performed. (See also article by June Christmas in this volume.)

(9) Research into the normal physiology of the female reproductive system and into the causes and treatment of gynecological diseases is grossly inadequate. For example, women have been suffering from menstrual cramps for a long, long time. Because doctors have regarded dysmenorrhea as "all in the mind," no attempts were made to determine the physiological basis for this painful condition. Even when research into prostaglandins suggested a relationship between these hormones and uterine contractions, no investigations into the use of anti-prostaglandin drugs for menstrual cramps appeared for many years. In fact the major research which led to the eventual marketing of these drugs was done by a woman family physician in her private office. The drug companies, private foundations and NIH, organizations which usually fund research so that the studies can be done on a large scale, refused to support her. The drug companies jumped on the bandwagon only after the results of her study showed they could make a fortune (Budoff, 1981).

Although the above examples delineate the poor medical care that women receive because of prevalent sexist ideas, I want to stress that men don't always get great medical care. Poor medical care is inherent in the current social system and sexism and racism only make it worse for everyone (InCAR, 1979). Men see that they don't get treated as badly as women and so are willing to accept what they get. In addition many of them accept the same lies about women's inferiority and don't struggle to improve health care for women.

Biological differences between men and women require the specific medical care for women involved in obstetrics and gynecology. However there are no genetic determinants of behaviour that are sex linked and predispose women to be depressed, unreliable, anxious, stupid and hysterical as most physicians expect. Women and men should be informed about what constitutes good medical care and struggle together on every level to achieve it.

BIBLIOGRAPHY

Budoff, Penny
   1980   *No more menstrual cramps and other good news.* New York, G. P. Putnam.

Buckman, Ronald and *The Women's Health Study*
   1981   *Association between intrauterine device and pelvic inflammatory disease.* Obstetrics and Gynecology, 57, (3), pp. 269–276.

Corea, Gena
   1977   *The hidden malpractice.* New York, Morrow.

Fruchter, Rachel, and Boyce, John, and Hunt, Marsha
   1980   *Missed opportunities for early diagnosis of Ca of the cervix.* American Journal of Public Health, 70, (4), pp. 418–420.

Gold, Julian, and Coles, Willard
   1979   *Restriction of funds for abortion: 18 months later.* American Journal of Public Health, 69, (9), pp. 929–930.

Howell, Mary
   1974   *What medical schools teach about women.* New England Journal of Medicine. August.

InCAR (International Committee Against Racism)
   1978   *The public Health of racism in the United States.* New York, InCAR.

InCAR
   1979   *Toward fascism: racism and the destruction of public health.* New York, InCAR.

Levanne, K. John, Jean and R.
   1973   *Alleged psychogenic disorders in women.* New England Journal of Medicine. February, pp. 288–290.

McCarthy, Eugene, and Finkel, Madelon
   1980   *Second consultant opinion for elective gynecologic surgery.* Obstetrics and Gynecology, 56, (4), pp. 403–410.

Symonds, Alexandra
   1980   *Women's liberation: effect on physician-patient relationship.* New York State Journal of Medicine, 80, (2), pp. 211–215.

Willson and Carrington
   1979   *Obstetrics and Gynecology, 6th Ed.* St. Louis, Mosby.

THE EFFECTS OF CHILDBEARING ON WOMEN'S MENTAL HEALTH: A CRITICAL REVIEW OF THE LITERATURE

Susan Bram, Ph.D.

A short time ago I was interviewed over the phone by the editor of a popular family magazine about my research on parenthood. After a few preliminary inquiries, she posed her central question: "Is it true that parents live longer than nonparents?" She explained, "Some of our readers are very distressed by the recent emphasis on the advantages of childlessness and want to know the health benefits of parenthood." She felt the media had been biased in favor of the childless and that the public deserved to know the "truth."

Perhaps the editor's question can best be understood as an attempt on her part to mollify her readers' anxieties about recent changes in fertility mores and to offer them the promise of a "reward" for what is admittedly a difficult occupation—parenting. It is particularly poignant that she turned to science for a "truth" that would support her point of view, however.

Of course, although scientists are interested in investigating such issues as the impact of parenthood on health and mental health, they are rarely in a position to give people support for their value systems, or to help them make such personal decisions as whether or not to have children. In fact, it has been nearly impossible to investigate the question posed until recently, because such a small minority of women have chosen not to have children without a predisposing physical problem influencing their choice. Census data for the last 25 years indicates that on the average at least 90% of women in the United States married at least once during their lifetime, and of these it is estimated that, prior to 1970, at least 90% had children (U.S. Bureau of the Census, 1979). Thus, it had been quite difficult to investigate the impact of having or not having a child on a woman's life, because there was little variability in the population as a whole.

However, today we are experiencing large-scale social changes that have begun to affect fertility patterns in such a way that there is a much greater diversity of reproductive phenomena available for examination. For example, although the opportunities have hardly been equitably distributed, there are over all more chances for women to obtain a higher education today, a larger percentage of women are entering the job market, even with small children at home, and more are engaged in professions previously limited to men (Hoffman, 1977). Furthermore, there is a later average age at first marriage, a higher rate of divorce, more single-headed households as well as smaller families, more single-child families, a later maternal age at first birth, and an increased rate of voluntary childlessness than in previous decades (U.S. Bureau of the Census 1979; Van Dusen and Sheldon, 1976). Although such changes should theoretically enhance our collective tolerance for diversity, in fact at present there are indications that we may be entering a period of increased divisiveness and polarization, especially among women who have participated in these changes and those who have not.

Unfortunately, scientists are not free from conflicts experienced by the public. Since research problems are generated by social conditions, shaped by available funds, and influenced in part by the race, class and gender of the researcher, it is possible that those "authorities" to whom people turn may unwittingly promulgate, rather than challenge, prevailing social myths. This is a particular danger in the study of reproduction in which personal interests and societal needs may be at odds, and when myths of the "joy of motherhood" and the sacred rite of childbearing abound (Bernard, 1975; Rich, 1976). Although most of the studies of childbearing have come from the field of medicine, psychology and psychiatry have often served as "handmaidens" in the process of transmitting cultural norms to the public in the form of "scientific fact." An underlying rationale for many of the studies has been that of "biological determinism."

We can hardly imagine a topic more vulnerable to a sociobiological analysis than the issue of childbearing. All societies require some reproduction of the species to survive, but who reproduces, how fast, and in what number are factors that vary according to the needs of the society at a given time. In our post-industrial society the pattern of childbearing has a great influence on the number of laborers and consumers, as well as dependents and those in need of "non-profit" services, such as education and health care. An example of the influence of government policy on reproductive norms is found in the World War II period. During the war, women were needed to work in the factories and were thus explicitly encouraged to reduce their family size and become fulltime workers. After World War II, women were then channeled back into the home, through the press, media and advertising, in order to make room in the labor market for returning vets. Psychologists participated in the social trend towards fulltime mothering by authoring childrearing manuals and conducting studies that emphasized the needs of children for a fulltime female caretaker and the pleasures of engaging in fulltime parenthood. Until the 1970's, with the rise of feminism, and the new critique of the social sciences, there had been a proliferation of such studies, which in various ways, confirmed the ideal of the "joys of motherhood." Perhaps today, at least according to the editor who interviewed me, we are erring on the other side by giving undue emphasis to the need for women to work outside the home, in order to contribute to the middle class life style, which now requires two incomes to survive.

My goal in this conference is to illustrate how assumptions rooted in biological determinism are utilized by medical, psychiatric, and psychological researchers to maintain control over the extent to which woman is defined primarily as a childbearer. To do this I will examine ways in which the assumption that childbearing is an essential aspect of female development is implicit in the mental health literature on childbearing and how this emphasis on biological determinism results in a serious neglect of the social conditions that are at least of equal importance in defining the experience of parenthood.

As an introduction to the review of the literature I would like to refer to two fundamental and related principles in psychological research that underlie much of the research to be covered, and contribute to the biases in the field: reductionism and functionalism. One major assumption, reductionism, is expressed in the "rule of parsimony" which is taught in most introductory college science classes. Stated simply, this rule means that the researcher should seek answers with the least degree of inference from data. However, in psychological research this often translates into a crude biological reductionism. For example, if it is observed that women experience depression following childbirth, it is simpler to attribute that to the apparent and measurable hormonal changes in the body than to more complex and subtler psychological experiences or situational changes that are much more difficult to pinpoint. The difficulty with both the rule of parsimony and biological reductionism in general in the study of human experience is evident: People experience their biology on an individual level; their subjective experience is shaped and influenced by the society and culture in which they live. Thus, any psychologist who wishes to capture human experience must be cognizant of biological "givens" as well as the social definitions of these "givens" at any point in history. Unfortunately, we are all too often limited in our investigations to one level of analysis, thus unduly constricting our view of the human condition. Of course, we also overlook the ways in which our own subjectivity and socio-cultural milieu influence us as scientists.

Another underpinning of much of psychological research is the functionalist approach, which has a long tradition in the field, originating with early evolutionary theory. According to this position, phenomena which survive the stresses of time are presumed to be, by definition, adaptive to the organism. In shorthand, this is often interpreted to mean, "what is, should be." In the study of reproduction and sex differences this principle has led many to believe that a woman's "biological potential = biological destiny = biological imperative." Since a woman *can* have a child, she *should* have one. A second extrapolation of this principle is that "what one person experiences, most people experience" and *vice versa*. This has resulted in such fallacies as the development of group norms for what is normal, healthy, and pleasurable for the majority. Thus, it is often implied that since most women have children, a woman who does not must be physically or psychologically unhealthy or unhappy. When scientists are confronted with data that challenge the functionalist assumption described here, they are often forced to develop new principles to explain the data. For example, if we discover that a woman who has children (and has fulfilled her biological potential) is unhappy with her role, it is easier to blame her (e.g., she is psychologically conflicted) than to re-examine the assumption that motherhood is adaptive for all women.

A brief review of the psychological and psychiatric literature of the past decade will illustrate how such basic assumptions, and derivatives of them, have resulted in a distorted view of the way in which a biological datum, the existence of distinct reproductive organs in men and women, can affect a psychological experience, thus feeding a social construction of reality that alienates men and women and separates physical and mental phenomena. New studies contributed by sociologists and social psychologists will also be described to demonstrate some alternative ways of investigating these issues.

Emergent Themes in the Literature on the Psychological Effects of Childbearing

Perhaps the most ubiquitous theme in the literature on psychological effects of childbearing is the notion that reproduction is woman's domain, not man's, and that it is "normal" for a woman to want and to have children, "abnormal" for her not to be so inclined. The study of reproduction is still focused primarily on women, almost as if scientists had not yet discovered the male role in conception.[1] Exclusion of men from research serves to maintain the exclusion of men from the delivery room and the nursery, despite the protests of feminist authors (Dinnerstein, 1977; Chodorow, 1978) and some modern fathers (Collins, 1981). This phenomenon is one of the clearest examples of how a strictly biological definition of the sexes has psychological and sociological ramifications that can result in a further separation, rather than integration, of male and female roles.

While most medical and psychiatric literature takes it for granted that women want to have children and are enthusiastic about pregnancy, childbirth and delivery, the view of expec-

---

[1]Ironically, according to anthropologists, it is in "primitive" societies that men are neglected in theories of fertility, because they have not comprehended the biological role of the male in conception. However, even in such cultures men have a more legitimized social role in the institution of childbearing through such rituals as the "couvade" (Mead and Newton, 1967), than do men in our society.

tant fathers is that they must be pitied or protected, since they presumably experience these events with fear, anxiety, or envy. It is only in recent years that psychologists and psychoanalysts have begun to take the psychology of fatherhood seriously and to study it in its full complexity (Fein, 1978; Lamb, 1976; Ross, 1979). In one attempt to integrate men into the psychoanalytic theory of parenthood, Benedek (1970) even describes the attitudes of "fatherliness" as "instinctive." She states that "fatherliness" is a derivative of the instinct for survival in which the psychological identification of the man with the father is achieved through an extension of the self in a (sic) son.

A second theme in the literature is the emphasis on "outcome." Woman's experience in childbearing is considered important to the extent that it has an impact on the fetus or the newborn. In fact, much of the data available on the psychology of childbearing come from studies designed with the goal of understanding the relationship between these two variables, rather than investigating ways to improve the health of the woman *per se* (Richardson and Guttmacher, 1967). Perhaps this is because much of the funding for women's health care studies derives from the principle that her medical status affects the health of future generations. Yet, ironically, as a result of this interest in offspring, the particular needs of the woman, usually the primary caretaker, are often overlooked. For example, until recently, there had been little attention paid to the experiences of women as obstetrical patients, their response to medical practices, such as episiotomies, anesthetics, or their preferences for modifications in health care delivery. (Hinds, 1981). Nor has anyone yet investigated the effect on women of being treated as the vehicle for childbearing, rather than as a participant in the process.

A third theme is the emphasis on biological aspects of reproduction, to the neglect of social conditions and psychological experiences. Throughout the many examples given below, for instance, it will be evident that childbearing is categorized according to medical markers, such as pregnancy, labor and delivery, or trimesters and stages of labor, rather than other subjectively meaningful phenomena, such as the time of conception, or the first missed menstrual period. Changes in social roles that accompany the transition to parenthood are usually overlooked in favor of the biological events.

One further and related theme in the literature is the emphasis on normative data, with little reference to the historical or social context of findings, or changes in reproductive patterns that affect the meaning of research results. One reflection of this narrow approach to research is seen in the design of studies. Many investigations of pregnancy are carried out without a comparison group of non-pregnant women and studies of childbearing have failed to include a control group of childless. Studies of the impact of pregnancy or childbearing on a woman's mental health have not always included measures taken before pregnancy, since respondents are often selected from clinic populations. Furthermore, medical research has been slow to include samples of newly emerging populations, such as the older mother or the middle class pregnant teenager. As a result, data on such topics as the optimal age for childbearing are often misleading (Daniels and Weingarten, 1979). Because of these problems with the research, any of the results given below must be interpreted with caution.

Research Results on Pregnancy, Childbirth, and the Post-Partum Period

The findings reported in the literature are fairly homogeneous in regard to the psychological experience of pregnancy. The common view is that pregnancy is a time of "normal crisis"—certainly a contradiction in terms!—marked by such "symptoms" as extreme sensitivity to stimuli, over-reactiveness, weeping, restlessness, nervousness. These "symptoms" are usually interpreted to mean that pregnant women are anxious, depressed, or neurotic. Bibring (1976) even described a phenomenon she called the "pseudo-borderline" syndrome of pregnancy, to distinguish the referrals sent by obstetricians to psychiatrists from the usual psychiatric patient with the diagnosis, "borderline."

Among the psychological issues observed by researchers during pregnancy are fear and ambivalence about the life changes accompanying pregnancy, such as the transition to motherhood, changes in the marriage and work situation, confusion about the changes in sexual image and appearance, and revival of thoughts about the mother-daughter relationship. Psychoanalysts have given particular attention to this latter theme and have emphasized the importance of childbearing in helping the adult woman resolve the oedipal conflict and achieve identification with her mother (Benedek, 1970; Bibring, 1976; Deutsch, 1973). Deutsch (1973) even went so far as to state that a woman who does not experience childbearing and pregnancy as a painful crisis is unhealthy.

In contrast, some authors emphasize the "heightened sense of well-being" during pregnancy (Hooke and Marks, 1962). In one of the few controlled studies of pregnancy, Paschall and Newton (1976) found that there was no evidence of greater neurosis among pregnant women than non-pregnant. They suggest that previous reports might have erred by interpreting behavior such as weeping as "neurotic." In fact, large-scale studies of psychiatric hospital admissions indicate that rates are *lower* among pregnant than nonpregnant or post-partum women (Kendell, Wainwright, Shannon, 1976).

In addition to the psychological symptoms, of course there are several common physical symptoms which can cause psychological distress, such as nausea and vomiting. While some authors suggest that these are merely reflections of ambivalence and "rejection of the child" (Nadelson, 1978), others have found that social factors may be more causal than psychological variables (Wolkind and Zajicek, 1978). For example, in one study, women with prolonged nausea during pregnancy were found to have received less emotional support from parents or husband.

Pain in childbirth is also attributed to psychological factors (Nadelson, 1978), although situational and social variables clearly play a role. Doering and Entwisle (1976) reported that women who are prepared in childbirth classes have a greater state of awareness during labor and report more satisfaction with the childbearing experience than those who are not prepared. Jones (1978) studied complications of labor and delivery and concluded that personality measures (e.g., MMPI) and measures of anxiety were much less effective than the Schedule of Recent Life Events (e.g., moving, changing job) in predicting problems such as high blood pressure during childbirth, low Apgar scores, and other obstetrical problems. Others have found similar results in their investigations of childbirth (Nuckolls,

Cassel, Kaplan, 1972; Rabkin, Struening, 1976).

Literature on the post-partum indicates that this is the most difficult phase of childbearing for a woman, although the interpretations of this finding vary. Review of data on psychiatric morbidity does indicate that there is a higher occurrence of some mild "emotional symptoms," especially depression and anxiety, during this time period for about one-quarter of women who deliver, although only a small percentage find their lives seriously impaired (Zajicek and Wolkind, 1978). There is over all more contact with psychiatric facilities during this time period than during other phases of childbearing and the hospital admission rate is higher among women post-partum than at other time periods measured (Kendell, Wainwright, Shannon, 1976). The estimated rate of actual "psychosis" during the post-partum varies, however, according to several aspects of the research, such as the population sampled, the definition of the syndrome, and the length of time considered to be "post-partum." Thus, the rate varies from about 1/1000 to 3/1000 (Herzog, 1976).

Although most clinicians nowadays prefer not to distinguish post-partum "psychosis" from other functional psychoses, others disagree. For example, one researcher has investigated the ways in which the sleep patterns among pregnant and post-partum women correlate with psychotic-like symptoms (Karacan, 1970). Others feel strongly that "childbirth is a genuine causal factor in the genesis of psychosis" (Kendell, et al., 1976, p. 297). Among those who espouse this point of view, explanations range from those that emphasize hormonal factors (e.g., estrogen, progesterone, thyroid, cortisol) to those that stress psychodynamic constructs, such as ambivalence towards separation (from the fetus), unresolved oedipal conflicts. Very few researchers have investigated the social conditions surrounding the transition to motherhood or the psychological experiences that accompany this developmental change. For example, becoming a mother does not only involve undergoing a physical transformation or reliving early psychological experiences. For most women today it involves a significant realignment of relationships with others, such as the spouse (Cowan, Cowan, Coie, Coie, 1978) and a disruption of an established work routine or career path. In addition, practical changes, such as a reduction in finances, impingement on one's space, and greatly diminished time, can alone upset one's psychological equilibrium.

It is thus clear that, although some of the literature reviewed yields evidence of the importance of social factors in determining the psychological experience of reproduction among women, the conclusions expressed by the authors tend to emphasize the biological variables. Even psychological theories stress constructs that are believed to be derivatives of biological forces. It is particularly striking that even though many studies have uncovered data that illustrates the importance of socioeconomic status, immigrant status, social and kin support, or working conditions on the psychological experience of pregnancy, childbirth and the postpartum period, the implications of these studies are often dismissed by the authors and are systematically ignored by practitioners (Bibring and Valenstein, 1976; Kendell *et al.*, 1976; Wolkind and Zajicek, 1978; Zajicek and Wolkind, 1978).

New Studies Which Stress Social Conditions

The new diversity of childbearing patterns of the last decade and a greater awareness of reproductive choices has led to the development of studies which investigate in more depth than previously the effects of childbearing on the mental health of women and men. Interestingly, however, most of the new research has come from sociology and social psychology, rather than psychiatry and psychology. Thus, the studies have taken a broader view of childbirth, emphasizing the institution of parenthood rather than simply the biological phenomenon.

Bernard (1973) was among the first social scientists to investigate ways in which being a mother or father affects mental health. Reviewing major social surveys of the 1950's, she discovered that, despite the fact that this was an era of high mystification of motherhood, there were strong indications that women were stressed by the role. For example, she found that in comparison to married men, married women (most of whom were mothers) were significantly more likely to report depression, severe neurotic symptoms, phobic tendencies, feelings of passivity, and psychosomatic symptoms. She concludes that the role of parent as traditionally defined has a differential, and debilitating effect on women. Other sociologists have documented in more detail specific ways in which the "transition to parenthood" affects women's self-esteem, feelings of autonomy, and relationship to spouse, at least during the initial stages (LeMasters, 1957; Rossi, 1968).

To understand the contribution of parenthood to the psychological stresses described, however, one must compare those who are parents with those who are not. Since most nonparents have often yielded data that contradict the traditional as- and often would have liked to have had children, they did not allow for a useful comparison group. However, the rising incidence of *voluntary* childlessness during the last decade has provided social scientists with an opportunity to do such comparisons. Studies which have contrasted parents with nonparents have often yielded data that contradicts the traditional assumption that parenthood is "bliss." For example, a large survey study by Campbell, Converse, and Rodgers (1976) comprising interviews of over 2000 people age 18–29 and 29–mid-70's, indicated that respondents who were young, female, married and childless were freer from stress than were those in other age, sex and marital categories. Furthermore, the older childless men were happier than other men and women studied. In general, the childless couples were found to be less conflicted as a group. Thus, it is evident that having children is not essential to individual or marital happiness.

Ongoing research by the author of this paper corroborates some of the findings of Campbell *et al.*, (1976). During the last ten years I have carried out an in-depth study of parents and nonparents at different points in the life cycle (Bram, 1978). Not only have I found that the voluntarily childless men and women were as happy and contented as the parents and parents-to-be, but there are some indications that their lifestyles are more self-actualizing and their marriages more egalitarian than those with children. Furthermore, it is significant that the response to the parenting experience varies according to both a developmental factor, age at first birth, and a social factor, the time period in which the family is initiated. For example, although parents who started their family in their early twenties in 1971–1972 reported psychological and financial stress during the transition to parenthood, those who delayed their first birth to a slightly later age and began families in 1978–1979 experienced even more turmoil. It is likely that two factors were at work. Probably the change in life style occasioned by the birth of a child at a later age requires a greater adjustment than earlier, particularly since most women have been working fulltime for several years previously. But, in addition, it seems that the general economic climate of the late 1970's made it a more difficult time to have and raise a child as well as to forfeit the income of the mother. A recent study by Weingarten and Daniels (1982) explores in more depth the experience of women who have children at a later age.

Recently, several longitudinal studies have been carried out which investigate pregnancy and childbearing over time to determine ways in which women and men respond to the process of change. Leifer (1980) conducted such a study by interviewing 19 white middle class women at several points throughout the pregnancy and post-partum period and found data that contradict earlier findings in the literature. For example, in contrast to the psychoanalytic view of pregnancy as a passive time in which early intrapsychic conflict is revived, she found it to be a psychologically active period in which much attention is devoted to developing the relationship with the new baby. She also discovered that a woman's success in coping with the pregnancy and childbirth experience was less a function of her personality and more a product of the strength of support she received from others around her, especially her husband. Gladieux' study of 26 married couples undergoing the first pregnancy (1978) also illustrates the importance of the social network in enhancing the woman's experience. She found that women with a close-knit network of relatives and close friends were more satisfied with their pregnancy and more optimistic about parenthood than were those who had few friends and were distant from their families. This was also true for men, inasmuch as those who had a network of friends who were themselves parents were more satisfied with each trimester of the pregnancy.

Yet another kind of study offers a critical look at the actual conditions surrounding health care for pregnancy and childbirth by using *in vivo* observations of hospitals and clinics (Arms, 1975; Shaw, 1974). Both researchers have pointed out ways in which the organization of childbearing can result in more pain, tension and dissatisfaction with the initial stages of parenthood. For example, the location of most childbearing in a hospital has led to an increasing reliance on technology, a separation of functions among personnel because of medical specialization, and a transformation of pregnancy into a "sickness" for many women. Since, as Rosengren (1961) demonstrated, pregnant women who engaged in the sick role often had longer labors, it is possible that the medicalization of this experience is actually exacerbating it (Parlee, 1976).

In summary, we are now just beginning to look at the issue of childbearing from a broader, more comprehensive perspective that allows us to see the ways in which a seemingly biological experience is shaped, not just by the personality and attitudes of the childbearer, but by the values and priorities of clinicians, physicians and scientists, and influenced by the economic and social context in which it takes place. Until we eliminate some of the sociobiological biases implicit in much of the research reported, we will not be able to begin to answer such questions as the one posed by the journalist about the longevity, let alone, the quality of life, of parents or nonparents. Nor will we be able to ensure safe or happy alternatives for those who do or do not have children.

REFERENCES

Anthony, E. J., Benedek, T. (Eds.)
1970   *Parenthood: Its psychology and psychopathology.* Boston Little, Brown and Co.

Arms, S.
1975   *Immaculate deception: A new look at women and childbirth in America.* Boston Houghton Mifflin.

Benedek, T.
1970   "Fatherhood and providing." In Anthony, E. J., Benedek, T., (Eds.) *Parenthood: Its psychology and psychopathology.* Boston Little, Brown & Co. pp. 167–185.

Benedek, T.
       "Motherhood and nurturing." In Anthony, E. J., Benedek, T., (Eds.), *Ibid.,* 153–166.

Bernard, J.
1973   *The future of marriage.* New York Bantam.

Bernard, J.
1975   *The future of motherhood.* New York Penguin.

Bibring, G., Valenstein, A. F.
1976   "Psychological aspects of pregnancy." *Clinical Obstetrics &*
June   *Gynecology,* 19 (2) 357–371.

Bram, S.
1978   "Through the looking glass: Voluntary childlessness as a mirror for contemporary changes in the meaning of parenthood." In Miller, W. B., Newman, L. F. (Eds.). *The first child and family formation.* Chapel Hill, North Carolina: Carolina Population Center, 368–391.

Campbell, A., Converse, P. E., Rodgers, W. L.
1976   *The quality of American life.* New York Russell Sage.

Chodorow, N.
1978   *The reproduction of mothering: Psychoanalysis and the sociology of gender.* Berkeley and Los Angeles University of California Press.

Collins, G.
1978   "Fathers get postpartum blues, too." *NYT,* April 6, 1981, B14.

Cowan, D., Cowan, P., Coie, L., Coie, J.
1978   "Becoming a family: The impact of a first child's birth on the couple's relationship." In Miller, W. B., Newman, L. F. (Eds.), *Ibid.,* 296–324.

Daniels, P., Weingarten, K.
1979 Spring "A new look at the medical risks in late childbearing," *Women & Health*, 4, (1), 5–36.

Deutsch, H.
1973 *Psychology of women*. Vol. II, New York Bantam.

Dinnerstein, D.
1977 *The Mermaid and the minotaur: Sexual arrangements and human malaise*. New York Harper.

Doering, S. G., Entwisle, D. R.
1976 December "Coping mechanisms during childbirth and postpartum sequelae," *Primary Care*, 3, (3), pp. 727–739.

Fein, R.
1978 "Consideration of men's experience and the birth of the first child." In Miller, W. B., Newman, L. F., (Eds.), *Ibid*.

Gladieux, J. S.
"Pregnancy—The transition of parenthood: Satisfaction with the pregnancy experience as a function of sex role conceptions, marital relationship, and social networks." In Miller, W. B., Newman, L. F. (Eds.), *Ibid*.

Herzog, A., Detre, T.
1976 April "Psychotic reactions associated with childbirth." *Diseases of the Nervous System*, 37, 4, 229–235.

Hinds, M. deC.
1981 Jan. 24 "Midwife births gaining wider acceptance." *New York Times* 48.

Hoffman, L. W.
1977 "Changes in family roles, socialization, and sex differences." *American Psychologist*, 32, 644–657.

Hooke, J. F., Marks, P. A.
1962 "MMPI-characteristics of pregnancy." *Journal of Clinical Psychology*, 18, 316–317.

Jones, A. C.
1978 Aug. "Life changes and psychological distress as predictors of pregnancy outcome." *Psychosomatic Medicine*. 40, 5, 402–412.

Karacan, I., Williams, R. L.
Oct. 1970 "Current advances in theory and practice relating to postpartum syndromes." *Psychiatry in Medicine* 1, 4, 307–328.

Kendell, R. E., Wainwright, A. H., Shannon, B.
1976 "The influence of childbirth on psychiatric morbidity," *Psychological Medicine*. 6, 297–302.

Kleinman, C. S.
1977 "Psychological processes during pregnancy." *Perspectives in Psychiatric Care*. IV, 175–178.

Lamb, M.
1976 (Ed.) *The Role of the Father in Child Development* New York Wilz.

Leifer, M.
1980 *Psychological effects of pregnancy: A study of first pregnancy.* New York Praeger.

LeMasters, E. E.
1957 "Parenthood as crisis." *Marriage and Family Living,* 19, 352–355.

Mead, M., Newton, N.
1967 "Cultural patterning of perinatal behavior." In Richardson, S. A., Guttmacher, A. F. (Eds.) *Childbearing—Its social and psychological aspects.* New York Williams and Wilkins, 142–244.

Measey, L. G.
Jan. "Psychiatric problems in obstetrics." *The Practitioner,* 220,
1978 1315, 120–122.

Nadelson, C. C.
1978 "Normal' and 'special' aspects of pregnancy: A psychological approach." In Notman, M. T., Nadelson, C. C. (Eds.), *The woman patient: Medical and psychological interfaces.* 1 *Sexual and reproductive aspects of woman's health care.* New York and London Plenum.

Nuckols, K. B., Cassel, J., Kaplan, B. A.
1972 "Psychosocial assets, life crises, and the prognosis of pregnancy." *American Journal of Epidemiology,* 95, 431–441.

Parlee, M. B.
Sept. "Social factors in the psychology of menstruation, birth and
1976 menopause." *Primary Care* 3, 3, 477–489.

Paschall, N., Newton, N.
Dec. "Personality factors and postpartum adjustment." *Primary Care,*
1976 3, (4), 741–750.

Rabkin, J. G., Struening, E. L.
1976 "Life events, stress and illness." *Science,* 194, 1013–1020.

Rich, A.
1976 *Of woman born.* New York Norton.

Richardson, S. A., Guttmacher, A. F. (Eds.)
1967 *Childbearing—Its social and psychological aspects.* New York Williams & Wilkins.

Rosengren, W. R.
1961 "Some social psychological aspects of delivery room difficulties." *J. Nervous & Mental Disease,* 132, 515–521.

Ross, J.
1979 "Fathering: A review of some psychoanalytic contributions on paternity." *International Journal of Psychoanalysis,* 60, 317–327.

Rossi, A.
1968 "Transition to parenthood." *Journal of Marriage and the Family,* 30, 26–39.

Shaw, N. S.
1974 *Forced labor: Maternity care in the United States.* Elmsford, New York Pergamon.

1979 United States Bureau of the Census, Current Population Reports, Series P-23, No. 70, "Perspective on American fertility."

Van Dusen, R. A., Sheldon, E. B.
1976 "The changing status of American women: A life cycle perspective." *American Psychologist,* 31, 106–116.

Weingarten, K. Daniels, P.
1982 *Sooner or later.* New York Norton.

Wolkind, S., Zajicek, E.
1978 "Psycho-social correlates of nausea and vomiting in pregnancy." *Journal of Psychosomatic Research,* 22, 1–5.

Zajicek, E., Wolkind S.
1978 "Emotional difficulties in married women during and after the first pregnancy." *British Journal of Medical Psychology,* 51, 379–385.

WOMENS' MENTAL HEALTH

Myra Fooden, Ph.D.

A concern for the status of women's mental health has been expressed amongst mental health professionals who deplored traditionally held psychiatric views. Similar concerns have been brought to public consciousness by the women's movement.

The issues relating to women's mental health may be expressed by asking three questions:

(1) What are the mental health problems generally attributed to women? Are these problems exclusively gender-based? If not, are they significantly more prevalent in women? Has this prevalence been scientifically validated?

(2) What are the suggested bases of gender-related mental health problems? Does the acceptance of gender-basing suggest the potency of one *etiology* over another? What evidence has been presented to substantiate one cause or another?

(3) What are the therapies suggested for the solving of gender-related mental health problems? How do these therapeutic modes relate to assumed causes?

While each of these questions could generate an extensive review, this discussion will be limited to demonstrating how any easily accepted assumption may contribute to insidious sexist/racist policies and practices.

Consider the first question; what are the mental health problems that women suffer? The most frequently discussed mental health problem of women is depression and this has been exploited by the media. While it has been acknowledged that depression may be a high order societal affliction (Beck et al, 1979), it has been more specifically characterized as a gender-related disorder. Substantiation comes from research evidence (Weissman, 1980) and practicing mental health professionals (Halas and Matteson, 1978; DeRosis and Pellegrino, 1976).

Although the word depression may convey a mood, a symptom or a clinical syndrome, how the word is being used is rarely defined clearly in much of the literature. By the time the more popular or quasi-scientific literature reports on depression, the descriptive criteria are so vague that any distinction is lost.

Epidemiological studies substantiate that clinical depression is more prevalent among women than men (Weissman, 1980; Weissman and Klerman, 1977). Weissman's finding that clinical depression occurs in 5.2 women per hundred, is based on research diagnostic criteria (RDC), a rigorously defined profile of depression. Beck (1979), however, reports that 15% of all adults, both sexes, between 18 and 74 years of age, may suffer depressive symptoms in any one year. His data are obviously based on different criteria than those used by Weissman, which serves to illustrate the difficulty in understanding the nature of depression. It appears that Weissman's criteria serves to eliminate non-*syndromatic* depression; it is unclear whether Beck's data refer to both clinical and non-*syndromatic* symptomatic depression, it certainly suggests a broader inclusion.

Symptomatic depression is often seen as being experienced in association with conflict situations. Professionals concerned with women's mental health are attempting to assess how the specific situational contexts of women's lives may result in symptomatic depression. For example, Wooley and Wooley, (1980) find depression a symptom associated with obesity in women. Thinness as a cultural value is an integral part of women's repertoires, and fatness is seen as a disorder. Preoccupation with weight loss, when unsuccessful, may lead women to avoid social contacts, avoid making important life plans and generate significant feelings of self-reproach. These behaviors are commonly seen as symptomatic depression along with feelings of helplessness and hopelessness, the emotional expression of depression. Gurman and Klein (1980) suggest that 25% of married women experience symptomatic depression. Concerned professionals seek to explain this phenomenon through analysis of the marital context and how the inherent situational factors present conflicts to women leading to symptomatic depression. The popularized "trapped housewife" and "empty nest syndrome" are two such attempted explanations, still being fostered although unsubstantiated.

Depression as mood change is probably felt by most people at some time. It is usually transient, self correcting and attributable to a variety of causes. It may be situationally prompted, thought generated or physiologically based. However, there has been a long history of emphasizing the relationship between mood depression and reproductive events, premenstrual and post-partum periods. The documenting by clinical observations, self report and folk wisdom has served to emphasize the incidence and to cast it as a women's depression problem, with the implication that women's physiology is inherently depression producing.

Anxiety is the second most frequently cited mental health problem in women. Like depression, anxiety is not a *single* concept, but is a class of disorders. Rejection of the notion of free-floating anxiety (Beck, 1976; Raimy, 1975) makes it necessary to more clearly specify anxiety disorders. Two predominant gender-related anxiety disorders are conversion hysteria and agarophobia (Chambless and Goldstein, 1980). While refuting the notion of women as hysterical personalities, these researchers cite studies supporting the incidence of psychogenic symptoms in women over men in ratios ranging from 2:1 to 4:1. Practitioners support, with anecdotal evidence, the chronicity of physical complaints by women, including chronic pain, jittery stomach, tiredness or weakness and shortness of breath. Agoraphobia, because of its prevalence in women, has been called "the housewives' illness." Agoraphobia has been described as a 'fear of the outside.' However, clinical evidence suggests that objects, situations and conditions generating the anxiety are highly individualized (Chambless and Goldstein, 1980; Raimy, 1975). Therefore, agoraphobia may be more properly viewed as a class of phobic responses. A further complicating factor is that many agoraphobics report experiencing depression or conversion symptoms. It is logical to suspect that the depressive symptoms are a result of the agoraphobic's inability to deal with the anxiety producing situation and the exceptionally limiting *functional outcome* of suffering the disorder.

In summary then the answer to the first question appears to be that depression and anxiety are more prevalent in women, but that these disorders must be more clearly specified. While Weissman (1980) has found a somewhat higher incidence of clinical depression in women, the experiencing of symptomatic depression as it has been described here, and the attention paid to depressive mood swings, have emphasized depression as a gender-related disorder. Certain situations are likely to cause symptomatic depression in either sex; however, it is suggested that women encounter more situations that cause the depressive responses than men do. Anxiety, the second disorder cited as gender-related, is more specifically defined in certain fear responses and life interfering behaviors as discussed. While the foregoing has documented the prevalence of gender-related problems, it has not discussed the reasons for the observations. An assumption that prevalence stems from a physiological base is not scientific. The same non scientific reasons hold when people believe that women are constitutionally weak mentally.

It is necessary to examine research evidence to ascertain relationships between incidence of problems and causes.

Attempts to explain the causes of women's mental health problems have been approached from physiological, psychological and socio/political points of view. Physiological explanations rely on genetic and endocrinological evidence and may be pursued by medically oriented researchers. Psychological explanations may range from intrapsychic psychodynamic theories to cognitive, learning and psychosocial theoretical frameworks, and be offered by psychiatrists, psychologists and social workers. Socio-political explanations rely on social theories and philosophic views of political/economic systems and have been generated by sociologists, anthropologists, social historians, social and feminist philosophers and by literary analysts who draw upon observations and reflections of the human condition. Attempts to understand the human psyche, and that of women in particular, have been made by researchers in all these disciplines. However, in a world where men are the standard, research data relating to women are scant, or, women have been measured as deviant from the existing standard. Also research from within a discipline tends to view data from within the field and rarely from a cross-disciplinary perspective. With this in mind let us examine the research and theoretical explanations.

Weissman (1980) finds that evidence for a biochemical basis of depression is meager. Genetic evidence comparing incidence within and between generations and comparison of *identical* and *fraternal* twins suggest some rationale for hereditability. These data are based in severe clinical syndromes and not the high-frequency symptomatic depression which accounts for the higher prevalence of gender-related depression. Also, there is no rationale for gender-specificity even amongst genetically related depressives. Weissman (1980) further found no studies using modern endocrinological methods of quantitative hormonal assessments in depression that would lend support to endocrinolcgical causation. While severe clinical depression in men and women may ultimately be found in some cases to have a biochemical base (Beck, 1979), the prevalent symptomatic depressive responses have no reason to be tied to the biology of women. Rather the stated relationship stems from an illogical inference, that because more women are depressed depression is 'naturally' related to biology.

Some researchers believe that eventually physiology will explain all behavior, however the complexity of the human mind points to psychological explanations as more productive. Psychological explanations for the prevalence of depression in women range from traditional *intrapsychic* theories to theories that are strongly social psychological in outlook. The former have been the object of growing criticism, while the latter have drawn increasing numbers of adherents amongst psychologists.

Kaplan and Yasinski (1980) argue that traditional psychotherapy assumes that the situational factors which bring men and women into therapy do not vary with gender. For both sexes, the source of the problem is thought to be totally within the person. In these views, the source of women's mental health problems would be assumed to be within her internal dynamics and not dependent on social situational factors, except as precipitating conditions. The traditional psychiatric view that when women were depressed in their roles as housewives, their failure to accept the role was due to some maladaption. A traditionalist view couched in the new feminism is Moulton's (1977) statement, ". . . social change brings about character problems that originate in the patient's childhood, but might not have brought her to treatment had she not been striving for new achievement" (p. 1). She believes that "all guilt and restrictions have been removed and that women have total opportunity," yet this has not created happiness. Views which rely on such traditional theoretical formulations promote sexist interpretations and become the basis for maintaining the status quo. Further such statements made by women with power and authority tend to be uncritically accepted by lay people, and can lead to further self-reproach in women.

Learning theory and social learning theories suggest that women's behaviors are derived from social conditioning and the incorporation of society's expectations. Typical of the first is that of learned helplessness (Seligman, 1974). It is argued that women learn to act helpless as a result of social conditioning. They act this way in the face of stress when there is a need to act assertively as in decision making or conflict resolution. Acting helpless is reinforced by the stress reduction when the source of stress is removed. Thus learned helplessness becomes the limiting factor in the response repertoire of women under conditions of stress. It is also characteristic of depressive symptomology.

An example of a theory of cognitive mapping of social behaviors is Bem's (1980) gender *schema* theory. Briefly, Bem suggests that children internalize a gender schema, and all incoming behavioral information is guided by the schema already internalized. Cultural biases, sex stereotyping practices, educational and childrearing practices, family systems and other socio-cultural systems provide the information. When ideas are transmitted early and consistently they fail to be evaluated, criticized and rejected by individuals. The source of negative emotional responses and destructive symptomatic behaviors evolves from distorted thinking about one's self and the world. Irrational thinking is related to the convictions, beliefs and notions about how one 'should' or 'must' think, believe or act (Raimy, 1975; Beck, 1976; Ellis, 1962). As incoming information conflicts with held beliefs, the inability to reject even nonproductive thoughts and behaviors causes negative emotional responses. The culture or the society is the origin of misconceptions; the human's propensity for irrational thinking is responsible for their continuance.

Further, acceptance and maintenance of a transmitted idea may be highly adaptive, at least in the short run. For example, Chambless and Goldstein (1980) point out that "for women a hysterical personality does not of itself constitute a pathology, since hysterical adjustment is congruent with sex-role stereotypes" (p. 114). Sex role stereotypic behaviors are most rewarded and reinforced, even though they may end up being non-productive for the individual. The expression of internal conflict through physical symptoms is often accepted and reinforced in women. Similarly girls and women who express fear of an outside situation are most often reinforced in their vulnerability. When these fears become entrenched phobic responses and are generalized to other outside situations, as in agoraphobia, it leads to a non-functional life-style. After these behaviors have been firmly established, society imposes negative evaluations on them. Since they are more prevalent in women, they are judged to be due to women's 'naturally' poor, ineffectual or weak attributes. In this way society perpetuates a genetic deterministic view. Women come to accept these premises and believe they have constitutionally weaker nervous systems.

Attempts to understand the source of women's mental health problems has been of concern to many feminist psychologists. The need for active research is recognized (Brodsky and Hare-Mustin, 1980). Only by understanding the mechanisms by which feelings and behaviors become part of women's repertoires, can an effective campaign of education against genetic deterministic views be mounted. It is especially critical to reach women whose own evaluative skills have not been well developed who will tend to accept apparent male authoritarian views. Traditional theories perceive the source of the problem within the person; but the problem is seen as amenable to change, if and only if, the person accepts the idea of being responsible (at fault) and accepts the proposed process of change. Theories which are more psychosocial suggest that the source may be found in a variety of situational factors and that the maintenance of the problem is due to related individual factors, including poor learning or opportunity for learning, poor evaluative processes, distortions and misconceptions in thinking based on limited information or experience. In this view the person still owns the problem, and is responsible for harboring it, but that through insight and understanding, learning of new responses and behavioral strategies and taking action the individual will be able to make changes and handle life situations more productively. The source of the problem is moved from a failure within the personality or character of the individual, in the traditional view, to the effect of situational contexts on unlearned or inexperienced persons. Both types of theories, however, converge in the attribution of responsibility for change, that is, to the individual's thinking and behavior.

A third approach to an explanation of women's mental health problems is expressed in socio-political terms. Society and its practices have been identified as the generating mechanism, with specific groups being victims. Many feminist writers, using a Marxist analytic framework have viewed women either as a minority group in a majority society or as a subjugated economic class. The emotional responses that women demonstrate are seen as expected, given the conditions of their lives and from which there appears little opportunity for escape. Radical change in society is proposed as the therapeutic approach. This presents the broadest contrast to the traditional psychological view, in which the individual's response is considered abnormal (Moulton, 1977).

Each of the different approaches to explanation has an implicit attitude towards the responsibility for having the problem. Because these attitudes are unexpressed they may be used to unfair advantage by those who wish to maintain the status quo. In the genetic deterministic model, women are not faulted for their behavior, they are told that they are victims of their endocrino*logical* system. This view is held by women as well as men. A carpenter attributed his forty-three year old wife's depression and agoraphobia to "her woman's problem, she must be starting." A divorced woman confided that she hoped not to get a certain family court judge on her alimony case, because she is reputedly unsympathetic in such cases. "She must be going through her changes," my informant said. Folk wisdom which ties emotional and behavioral responses to female physiology is very pervasive in spite of clinical or scientific evidence. It appears less nasty (actually more condescending) to blame a fact of nature than to attribute it to personal responsibility. For women themselves it is less onerous to blame their physiology than to accept the responsibility for seeking alternative explanations. Let me hasten to add that women in this male dominated society have been well educated and indoctrinated to live by the acceptance of their physiological weakness.

An interesting contradiction occurs in the case of obesity. Cultural pressures exert more negative effect on women than on men for being overweight. Despite the fact that there is no clear research evidence of the causes of excess weight, except possibly an inherited constitutional predisposition, weight reduction programs proliferate (Wooley and Wooley, 1980). Women are encouraged by medical practitioners (fat doctors) and pressured by societal expectations to pursue what may be an unreachable goal, an ideal figure or specific weight level. Weight reduction programs have a dismal rate of failure. Therefore, what may actually be determined by a physiologically based characteristic, a specific body weight/structure ratio is denied, and women are subjected to constant failure, which they and others attribute to personal inadequacy (Wooley and Wooley, 1980). Only when one has seen the terrible results in young women of eating disorders like Anorexia Nervosa and Bulimia, can one understand the insidious effects of a cultural pressure which becomes distorted by those developmentally vulnerable.

Responsibility for negative emotion and behavior (they act ugly because they feel ugly) is seen by socio/economic/political feminists to be rooted in those systems of which women are victims. The "victim" view creates a problem in that it tends to remove all responsibility from the individual as in the genetic deterministic model. This leads to an interesting paradox. Both the genetic deterministic model and the socio/political explanation remove responsibility for feelings and behavior from the individual and place them on a system, yet with entirely opposing expectations. In the first case the system is biological, a largely immutable one, and the implicit message is that one must accept the biologically given and live with it. In the second case the system is political/economic, a socially created system which is deemed to be more amenable to change. That of course, supposes that those who hold the political/economic power are sympathetic to the idea of change. While the ideal would be radical change advocated by social political activists, with regards to women's status, how realistic is the expectation that changes which would serve to ameliorate the conditions of large groups of women can occur? Take, for example, the issue of equal employment and equal wages for employment, which has been a strong focus of the feminist movement. During the last twenty years, the period of most active feminism and economic growth in the country, women's income dropped from 61% of men's to 57%. Adult unemployment among women rose from 9% higher than men to 43% higher and top jobs held by women fell from 5.5% to 4%, although employment of women rose from 25% of the labor force to 41% (Thurow, 1980). Thus it seems that while the collective actions of women have benefitted a few individuals, it has left more of them in disadvantaged positions. This is the very result that the genetic deterministic model has perpetuated and against which feminists have been most critical.

Other comments on the issue of responsibility have been made by psychologists and sociologists concerned with the effect on women's lives. Datan (1980) a psychologist says, "The professor, the feminist and the mother all distrust those aspects of feminism which displace the source of personal unhappiness from the individual on to social processes. They are all wary of any strategies which help the individual evade responsibility . . ." (p. 67). While Datan does not explicitly state support of a traditionalist view, she does reflect that view of many mental health professionals that the responsibility for emotional/behavioral change is within the individual, the socio-political system not withstanding.

Bernard (1981) also sees the victimization perspective as unproductive in assisting women to deal with the reality of their lives. "I take the misogyny of the male world as a given . . . . But exclusive emphasis on the victimization perspective seems to have a deleterious effect on female self-image and self-esteem," . . . (p. 31). It is logical to see that when emphasis is placed on victimization, it can lead to negative self-evaluation for allowing it, especially when a few token women are held up as successful in the world as it is.

The process of devictimization is certainly not agreed upon, given the harmful effects of established socio-/economic/political systems. Battle-Sister (reported in Bernard, 1981) a radical feminist believes that creation by women of counter institutions may serve only as "a placebo, a comforting cocoon, a retreat from the male world" and as such may be counterproductive since it would be merely part of a counter culture without a power base.

A different view is expressed by Berit Ås, another sociologist whose work was also cited by Bernard, who believes that the creation of a female culture would serve to create a real and informed picture of women's experience, could offer therapeutic tools for self-esteem and consciousness in women and might make women's problems politically potent. One can intuitively appreciate and agree with the first two of the outcomes, to understand women's lives and to offer support to women. This has already been part of the effectiveness of the current women's movement. However, the achievement of the last goal, to affect political change is questionable. This hoped for goal depends on the openness and acceptance in the majority male society and its voluntary willingness to change. To believe this, it is necessary to view misogyny, not as a deliberate strategy for maintaining power, but as a result of uninformed idealogical evolution, of which men are as much victimized as women.

This issue of responsibility for cause or responsibility for change in the social/ethical sense is not easily resolved. This is especially pointed when women are still being victimized on the simple level of their physiology. However, the fact that serious attention has been given to this issue by those cited as well as by the writer suggests it is a critical issue in relation to the effect on women's emotions and behavior. For mental health professionals who are concerned with helping individual women lead productive lives in the present situation, a decision has been made. While they have acknowledged the faults of the social/political system and accepted that this is the source of negative emotional responses common in women, the task at hand is to teach women how to survive and work toward change. This approach has been called feminist therapy. Thus, the moderate feminist objectives have been incorporated into a therapeutic view directed towards alleviating the mental health problems prevalent in women.

Yet, radical feminists see this merely as another plot of the woman-hating male society. Daly (1979) believes that women who became feminist therapists were seduced by the male society into being victimizers of other women because the establishment recognized their potential as deviants. While she agrees that women's collectives as crisis counseling centers may be needed, she is critical of institutionalized therapy, that is a system which keeps its clients dependent on the system. She sees this as incompatible with the full sense of courage of self-acceptance, that is, accepting responsibility for one's process of change. It is necessary to examine the premises of feminist therapy to evaluate the validity of Daly's criticism, which shall be done in the following section.

The foregoing section has illustrated that the question of the cause of women's mental health problems is most difficult to deal with. As the discussion has moved from a simplistic physiological question to more psycho/social explanations, the ramifications are exceptionally complex. While it is impossible to define cause, it may be interesting to look at approaches to suggested therapies.

The presence of a problem usually provides motivation for some concerned individuals to seek a cure. Therapy is a process expected to induce change in individuals to whom it is directed, while changes in systems of society which engender unhealthy responses is seen as social change.

Therapies which are directed at individuals include drug therapies, which purport to mediate behaviors to allow persons to become more functional; and psychotherapies which give individuals insight, understanding and strategies for behavioral change which allow them to make choices and develop expanded repertoires with which to function adequately.

Psychopharmacology is a field which has expanded significantly within the last twenty years. While the efficacy of tricyclic anti-depressant drugs has been documented (Weissman, 1980; Beck, 1979), the warning that different types and clinical pictures of depression may require differential treatment modalities (Beck, 1979) has not been heeded. Several studies confirm the frequency and higher dosages of drug treatment prescribed for women than for comparably diagnosed men (Brodsky and Hare-Mustin, 1980). Such overprescription may lead to drug abuse. Concerned therapists are critical of the tendency to treat women's mental health problems indiscriminately with drugs. Drug dependence as well as dependence on the pharmacological approach in lieu of problem solving is the real and present danger.

Women are also more frequent consumers of psychotherapy than men. Several rationales have been proposed as explanations of this observation. Women suffer more mental health problems of depression and anxiety, their more general complaints are more often viewed as psychological rather than physiological and women themselves are more willing to admit the need for help. Which of these reasons appears to account for the higher consumption of psychotherapy by women is yet difficult to determine, since research into this question has just begun (Brodsky and Hare-Mustin, 1980).

Not only are questions of consumption of psychotherapy of interest but many others relating to the therapeutic process and women. For example researchers are questioning the effect of the gender of the therapist and the client; of the gender-related attitudes of both therapist and client on the process and outcome of therapy; of variables such as age, marital status, client personality interacting with gender; of the choice of approach to therapy and of the development of criteria for assessing outcomes in therapy. While therapy is a difficult process to define, assess and measure, those especially concerned with women's mental health recognize the need to explore all of these factors if effective services are to be provided.

In spite of the state of the art, new therapies are constantly being developed. One which has garnered much interest amongst women in the last decade is feminist therapy. This therapy grew out of the concerns of feminist oriented therapists, to provide therapeutic services to women which were free of prior idealogies, of the dynamic interactions common to non-feminist practictioners (Lerman, 1980).

Even amongst practicing therapists questions are raised; what is feminist therapy? Who are feminist therapists; how does it differ from other therapies and is it necessary to have yet another therapy just for women?

Maracek and Kravetz (1977) define feminist therapy as follows: activism for social/political change as well as personal; a commitment to feminist principles of political social and economic equality between the sexes; understanding women's problems mainly within a socio/political framework; fostering development of feelings of personal strength, acceptance of self, recognition of self as equal in all interpersonal situations, and fostering feelings of trust of self and other women. Most feminist therapists would agree with this definition. It is compatible with the rhetoric of feminist philosophy.

However, there is always a problem in translating philosophy into action. While there are many who believe in feminism, there are different conceptions of how these beliefs translate into behaviors in the real world (Kolbenschlag, 1979; Daly, 1979; Hall, 1979; Mandler and Rush, 1975; Habib and Landgraf, 1977; McBride, 1976; Williams, 1976). For example, Daly sees the existence of a 'hag' society as the medium for true self-acceptance and the countering of male institutional systems; McBride, a married feminist dwells on the mutuality of shared equality in interpersonal experiences.

As a movement in growth, feminist convictions are being defined and refined. The movement has shifted from its earlier 'core' ideal of a radical critique of family and *phallocentric* heterosexuality to concerns with the real social, economic and political power issues facing women (Haber, 1980). Many messages are sent; how anyone selects, listens and integrates all or any part of these messages with personal experience will determine a philosophy.

This is important, for I believe, a therapeutic intervention emerges from a personal philosophy about the human condition, combined with a theory about human functioning and a methodology for implementing therapeutic change. Feminist therapists are those who subscribe to the philosophies of feminism, but there are no criteria for being a feminist therapist. Because the therapy rose from the consciousness of a select group within the mental health professions, they identify with those who hold similar beliefs and convictions. They were trained therapists first, from diverse methodological approaches and feminist therapists by the incorporation of a philosophical idea.

It is important to note however, that feminist therapists share the enviable position of being part of the privileged elite. Bernard's (1981) insightful work on women's world suggests that the experiences of the elite group are quite different than the experiences of others in that world; a world often separated by age, education, social status by affiliation, color, race, ethnicity and financial advantage.

While social historians have provided insight into the process which has led to current social inequalities (Chafe, 1977), the effects of characteristics such as social class, race and ethnic values on the therapeutic relationship are not well understood (Jones, 1981, 1974; Wilkerson, 1980).

The current feminist movement attempts to provide a new

way of interpreting the human experience for women in the context of today's society, but women's experience is not a single one shared by all women in the same way. Therefore, the mechanisms for change will not be the same for all.

The goals of feminist therapy are: to provide women with insight and understanding that the sources of their unproductive feelings result from conflicts with socially learned attitudes and behaviors and are not a result of personal inadequacy or biological limitation; to teach women strategies for personal change within the context of their social world; and perhaps to provide them with the skills and knowledge for effecting impact on others in their environment. To the extent that the existence of feminist therapy has served to develop consciousness to the needs of women and concerted action to meet those needs, and to the elimination of stereotyping of women (Sherman Koufacos and Kenworthy, 1978) it is performing a necessary and effective function, which could not have been carried as far by existing systems. Probably the most important factors in feminist therapy's favor are that methodological approaches are not yet fixed and that known dynamics of therapeutic situations are open to examination.

For example, age is a characteristic which may mitigate against the application of feminist principles in counseling, even for white middle class women (Haber, 1980). We know little about the ongoing accomodation between individuals and the environment across the life span. There are few studies of adult development, especially dealing with women's lives. Lowenthal (1975), using several different age groups of women found personal changes across life stages, especially in life satisfaction. For individuals with limited life spans, today's doctrine may be next year's downfall. Changes in personal sitautions, in attitudes and behaviors may be incompatible with conditions of decreased economic/job opportunities, with changing social interaction patterns, with changes in living conditions and with breakdown in support systems, both personal and societal.

Women in transition whose consciousness has been raised and who are looking to express personal change may be left in conflict when they discover the outside world hostile to their new changes. Current increases in crimes against women, in behavioral symptomology (alcohol or drug abuse) and other crises in personal lives may result from being caught between their expectations and reality. Feminist therapists and psychologists need to create a theory of personal life and adult development in women without preconceived notions. Only in this way will there be help for women in transition and beyond, especially those for whom radical life changes may not be realistically based.

In addition to age, other attributes of ethnic, racial and cultural values must be critically evaluated within the context of feminist therapy. Wilkerson (1980) emphasizes the need to construct reality-based models of the experiences of poor and minority women; to investigate the racial/ethnic beliefs held by therapists and to understand multi-cultural and multi-lingual differences and their effects in practice with minority women. For feminist therapists, who are mainly white, middle class and educationally advantaged, these are crucial in order to preserve the ideal of equality and effectiveness.

Feminist therapy and therapists have already given thoughtful consideration to many of the dynamics that are operant during therapeutic interactions. For example, there is a power inequity inherent in all therapeutic dyad situations; the therapist is seen as the knowledgable, paid authority (Brodsky, 1980). Being particularly sensitive to this, feminist therapists used self-disclosure, the sharing of women-to-women experiences and support in client decision-making to lessen the impact of therapist power.

Modeling is another dynamic in therapeutic relationships, where the client assumes the ideas and behaviors of the therapist. This has been considered a positive factor in feminist therapy, to model after women who are assertive and have self-esteem. While it is true that therapists might be seen as models of 'women who have made it,' such a dynamic might have negative effect. If the therapist is seen as the ideal model to copy, but actual attainment of status eludes accomplishment by the client, negative self-evaluation might result (Daly, 1979). It is necessary to know more about a client's inferences in role modeling, degrees of similarity between clients and models and the specific context of therapy to understand the impact.

The encouragement of women to express their long repressed anger was considered important in feminist therapy theory. Yet consideration of what was to be done with the anger was left unexpressed, or unexplored. Dissipation by pillow-throwing, bat beating or primal screaming are all outlets. Displacement onto other powerless persons was also a possible outcome, certainly not considered appropriate. The danger of creating "anger-junkies" (McBride, 1977), people hung up on expressing their anger also exists. This would, of course, be personally non-functional. The anger needs to be mobilized into productive energy for social change. Yet there is little evidence that we have systematic strategies for mobilizing that anger, or even that it is feasible in the context of therapy. The reason for the anger may also be critical; for middle class women, it may well have been generated by gender stereotyping, yet the same anger may stem from an entirely different process in poor and minority women. Chafe (1977), warns that a full analysis of how sex, race, class, ethnicity and regional background have interacted in American Society has not yet been done, and it is therefore unwise to generalize with a group called 'women.' Carrying this one step further it is unwise to generalize to the specific sources of their emotions.

As noted previously Maracek and Kravetz' (1977) definition of feminist therapy included activism for social/political change as well as personal, as one of the features. Whether expectation of women seeking therapy to work for social change is appropriate must be considered. Women seek counseling for personal problems. Providing them with understanding that the source of negative feelings may be an effect of the social/political impact on their lives is important. Activism by the therapist is highly appropriate, but whether clients' personal change must include social activism bears examination.

For consideration of this question, Chafe (1977) provides some interesting insights. He believes that, whatever the contributing social forces to social change might be, the crucial component is collective activity. He further describes two kinds of behavior; first aggregate behavior which refers to activities women engage in and appear to be based on gender, but which do not reflect group conciousness or planning; and collective behaviors which grow out of a sense of group purpose and reflect group intent. Chafe further states that just because people are treated as a group by others, does not mean they perceive themselves as a group; and when referring to women a sense of conscious group-ness must precede their perception of being collectively discriminated against and victimized.

Chafe has suggested that there are four stages of group behavior among women. The first refers to the activities women perform in society similar to other women, but with no sense of it being because they are 'women', that is, no sense of groupness at all. The second is where women act in awareness of their common indentity, but without group planning; this is the assumption of women's place or women's work. These are aggregate but not collective behaviors. A third stage described by Chafe is the consciousness of group identity and the planning of activity related to groupness. This is seen in charitable groups, self-help groups and sharing groups. Consciousness-raising, an important activity early in the feminist movement appeared to be instrumental in moving women from stage two to stage three and the many self-help groups emerging from this activity was evidence of its success. The fourth stage is collective activity in the form of a protest or social movement against the dominant society and the status quo. It is a shared consciousness of anger and protest against oppression and victimization.

In feminist theory, on which feminist therapy is based, change begins with the self, with the ideal that if enough equality could be effected in the relationships of some who were leaders, they could serve as models for others to do the same. The swell of such a group could provide the groupness of protest and exert pressure for social change in the extenal world. And to some extent this was the effectiveness of the early movement, for the "innovators and early learners," as Bernard (1981) has described them. For the vast majority of women who seek therapy, the development from stage two of which some vague message has made them discontent, to stage three of group consciousness, especially that which involves self-help activities, may be a giant step. Of course, stage three consciousness has also produced groupness which resulted in counter feminist ideals such as the 'Total woman' concept. This concept reinforces the dominant culture's view of biological determinism. The effectiveness of the feminist therapist as a model of anti-biological determinism cannot be underestimated.

The problem exists that women have been traditionally separated in American society and indeed live most of their lives more in constant contact with men than with each other. This mitigates against the effective development and maintenance of a collective consciousness of anger and protest that could be mobilized against the dominant society. One further point is the climate in society created by the political, economic and social forces must be nurturant for social protest to become social change. It appears that since the 1960's and early 1970's when the movement became vital, the climate has changed. Perhaps we need to understand more about the mutual effect of personal change and societal change and that the most effective tool we have now is the consciousness of collective behaviors that allows women to create stronger and stronger support groups within the women's world. Perhaps then feminist therapy's objective might be to form groups, not only for self-actualization, which it has already done, but for mobilization of anger to build stronger support groups. There has been some activity directed at this, but it has been mainly within the context of human services, i.e. groups against rape, pornography, battered wives and so on. It needs yet to be extended politically and economically.

We started by asking about cures, what is a cure for a social ill? A therapy is not a cure; it is a change in some system amenable to change. Can feminist therapy change the system responsible for creating the environment which leads to negative emotional/behavioral responses in women? Hardly. What it can do is provide women with the environment, not only for self actualization as an individual activity, but group actualization as a necessity for survival. Both feminist therapists and researchers will have the unique opportunity to study, collect, analyze and interpret the real experiences of women across class/ethnic/racial and religious backgrounds. Each event, statement, theory or position that perpetuates biological determinism must be exposed, and only women whose collective consciousness has been raised can do that.

REFERENCES

Beck, A. T.
1976 *Cognitive Therapy and the Emotional Disorders.* New York, International Universities Press.

Beck, A. T., Rush, A. J., Shaw, B. F. and Emery, G.
1979 *Cognitive Therapy of Depression.* New York, Guilford Press.

Bem, S.
1980 Gender schema theory: a cognitive account of sex typing. Invited address at the American Psychological Assn. meeting, Montreal.

Bernard, J.
1981 *The Female World.* New York, Free Press.

Brodsky, A. M. and Hare-Mustin, R. T.
1980 *Women and Psychotherapy.* New York, Guilford Press.

Brodsky, A. M. and Hare-Mustin, R. T. Women and psychotherapy: priorities for research. In Brodsky and Hare-Mustin (eds.) *Women and Psychotherapy.*

Brodsky, A. M.
1980 Feminist therapy: a review of the 70's, future directions for the 80's. Paper presented at the American Psychological Association meeting, Montreal.

Chafe, W. H.
1977 *Women and Equality:* changing patterns in American culture. New York, Oxford Press.

Chambless, D. L. and Goldstein, A. J.
1980 Anxieties: agoraphobia and hysteria. In Brodsky and Hare-Mustin (eds.).

Daly, M.
1979 *Gyn/Ecology.* Boston, Beacon Press.

Datan, N.
Spring Days of our lives, Journal of Mind and Behavior, 1 (1).
1980

DeRosis, H. and Pellegrino, V.
1976 *The Book of Hope.* New York, MacMillan.

Ellis, A.
1962 *Reason and Emotion in Psychotherapy.* New York, Lyle Stuart.

Gurman, A. S. and Klein, M. H.
1980 Marital and Family Conflicts. In Brodsky and Hare-Mustin (eds.).

Haber, B.
1980 Is personal life still a political issue? Feminist Studies, 5 (3).

Habib, M. and Landgraf, B. J.
Nov. Women helping women. Social Work, 22 (6).
1977

Halas, C. and Matteson, R.
1978 *I've Done So Well, Why Do I Feel So Bad.* New York, MacMillan.

Hall, C. M.
1979 *Woman Unliberated.* Washington, D.C. Hemisphere Publishing Corp.

Jones, E.
1978 Effects of race on psychotherapy process and outcome: an exploratory investigation. Psychotherapy: Theory Research and Practice, 15 (3).

Jones, E.
1974 Social class and psychotherapy: a critical review of research. Psychiatry, 37 (2).

Kaplan, A. G. and Yasinski, L.
1980 Psychodynamic perspectives. In Brodsky and Hare-Mustin (eds.).

Kolbenschlag, M.
1979 *Kiss Sleeping Beauty Goodbye.* New York, Doubleday.

Lerman, H.
1980 Remarks made as discussant in Symposium on From feminist mystique to feminist therapy presented at the American Psychological Assn. meeting, Montreal.

Lowenthal, M. F., Thurnher, M. and Chiriboga, D.
1975 *Four Stages of Life.* San Francisco, Jossey Bass.

Mandler, A. V. and Rush, A. K.
1975 *Feminism as Psychotherapy.* New York, Random House.

Maracek, J. and Kravetz, D.
Nov. Women and mental health. Psychiatry, 40 (4)
1977

McBride, A.
1976 *A Married Feminist.* New York, Harper and Row.

Moulton, R.
Jan Some effects of the new feminism. American Journal of Psy-
1977 chiatry, 134 (1).

Raimy, V.
1975 *Misunderstandings of the Self.* San Francisco, Jossey-Bass.

Seligman, M. E.
1975 *Helplessness: on depression, development and death.* San Francisco, W. H. Freeman.

Sherman, J., Koufacos, C. and Kenworthy, J.
1978 Therapists: their attitudes and information about women. Psychology of Women Quarterly, 2 (4).

Thurow, L. C.
1980   *The Zero-Sum Society.* New York, Basic Books.

Weissman M. M. and Klerman, G. L.
1977   Sex differences and the epidemiology of depression. Archives of General Psychiatry 34.

Weissman, M. M.
1980   Depression. In Brodsky and Hare-Mustin (eds.).

Wilkerson, D.
1980   Minority women: social-cultural issues. In Brodsky and Hare-Mustin (eds.).

Williams, E. F.
1976   *Notes of a Feminist Therapist.* New York, Praeger.

Wooley, S. C. and Wooley, O. W.
1980   Eating disorders. In Brodsky and Hare-Mustin (eds.).

THE HEALTH OF OLDER WOMEN IN OUR SOCIETY

Georgine M. Vroman, Ph.D.

This essay will deal with the present generation of older women, that is with those women who are now over 60 years old. This age limitation is an arbitrary one, but it does have some social significance. For instance, the age of 60 conveys certain Social Security benefits. Specifically, a widow or widower of someone entitled to Social Security payments will not lose their share if they remarry after the age of 60. This group of older women includes both the socalled "active" old and those for whom aging has brought some kind of disability limiting their activity, whether this disability be of a physical, mental or social nature, or a combination thereof. Furthermore, personal exposure and the kind of literature surveyed have led to an emphasis on women of working- and middle-class backgrounds. This paper does not in all respects apply to women at the extremes of the socio-economic scale. Neither will it address in detail the specific situation of women whose membership in a "racial" or ethnic minority is considered as their most important characteristic, by others or themselves.

The position of older women in our society is, in general, not only highly complex, it leaves much to be desired. Consider the following points. The condition of our elderly in many ways is determined by a combination of prejudices, misconceptions and forms of injustice, although the impact will, of course, vary for the individual case. Older men and women both suffer from this form of discrimination. For the older women, however, there is an additional combination of prejudicial attitudes and specific circumstances related to the very fact that they are women. In the first place, there are those factors that depend on their "biological" sex, that is, on their specific female physiology and anatomy. It cannot be denied that women's bodies are different from those of men, and possibly in more respects than in their different reproductive functions. Women's bodies may indeed age in ways that differ from the ways men's bodies age. However, this finding may not be true in every respect. People may be too hasty in assuming that such differences are proven on scientific grounds, while, in reality, at least some have no basis in fact and only reflect people's expectations and preconceived ideas. Conferences such as the present one can contribute significantly to clarify such distinctions. In the second place, older women differ from older men in their social situation. Older women are in the final stages of a lifetime, spent as female children first, then as adolescents, and finally as adults in a society that treats its male and female members very differently. Although this remains true for women at the present time, and will continue to be the case in the foreseeable future, each generation shares specific experiences that will differ from those of previous and future generations. For those of us who were young adults in the 'fifties, it is not easy to imagine what middle age will be like for our sons and daughters and how their respective situations in life will develop.

During this conference we will, as we did in the previous ones, make an attempt to distinguish between the physiological and social aspects of being a woman, and focus on ways where the two interact. In this paper we will investigate how some of these factors express themselves in the health of women over 60. The task of unravelling the tight web of physiological and social factors is difficult and may never be completed. Still, we must try, because on the outcome depends the potential for present and future change.

To begin, how should we define health? It is not a simple concept and it has held different meanings for different peoples at different times. Some criteria could be called "positive" such as soundness in body and mind. Other criteria stress the necessity that certain conditions be absent, namely disease and injury. In 1946, the World Health Organization defined health as follows: "a state of complete physical, mental and social well-being and not merely the absence of disease and injury." We will take this definition as our guide. For older women we must consider, on the one hand, the effect of such specific diseases as cancer of the reproductive organs and the breast. We have to add cardiovascular disease, respiratory disease and cerebrovascular disease, to which older women and men are particularly susceptible. On the other hand, we have to bear in mind the stress, mental anguish and various forms of alienation that must follow from the social isolation of having become a superfluous member of a society, preoccupied with youth and with work-for-pay as a yardstick of one's social worth. Men and women, both, suffer from such isolation in old age, but, as we will discuss, there are some important differences in the ways it affects them.

The degree of isolation depends to a large extent on the ability and opportunity to participate, to keep functioning. It is especially important how older people themselves experience this capacity. For some this continued participation may be the opportunity to travel or to partake of the many cultural offerings available in large metropolitan centers such as New York. For others it may mean a job, or possibly, involvement in a daughter's family. It is important to feel needed, useful, still part of what is going on in the outside world. One's sense of well-being, of self esteem, depends on this, no matter what age group one belongs to. This becomes particularly essential when one's circumstances in life make such participation difficult, as it is for the retired, the disabled and the homebound. Traditionally, women, locked inside their roles as housewives or denied full opportunities as workers have suffered from a sense of isolation and, for many, it has profoundly affected their self esteem. The following quote should clarify the ways this active participation relates to people's health. Dubos, the microbiologist and environmentalist, has written: "Health in the case of human beings means more than a state in which the organism has become physically suited to the physiochemical conditions through passive mechanisms; it demands that the personality be able to express itself creatively" (1965, xviii). In the present context we interpret creativity as active functioning and using one's resources imaginatively as participant in one's total environment. As mentioned, how each individual experiences this need to function, and how it is expressed, depends on many factors, individual, as well as historical and cultural. But among the most important determinants must surely be counted a person's own life history and the stage of the life cycle (*ibid*, 348 ff.). Butler and Lewis (1974) used this concept in their psychotherapy for depressed elderly men and women. These people were helped to explore their own life history and experiences in a serious and thorough manner. They responded positively to the realization that they were "successful survivors" and not "has-beens" and they discovered that their experience could be of interest and useful to others, younger than themselves.

In what ways does the aging process affect women's health in particular? We are, of course, using the WHO's definition of health, mentioned earlier. Let us first consider physical (physiological) health. It is important to realize that chronological age is not a good yardstick for measuring the degree to which physiological aging has occurred, in the individual case. It is also important to emphasize the need to investigate in what respects the physiological aging process differs for women and men. The number of years one has lived are no reliable indication of the extent to which one's body has aged, physiologically. Physiological aging, the wear and tear that occurs over time, is a slow process that affects different organ systems of the body at different stages of life and at different rates. A good example is the menopause, the cessation of menstruation, which usually occurs between the ages of 45 and 50. Although it heralds the end of the particular woman's ability to reproduce, it does not mean that, henceforth, she will no longer be able to share sexual enjoyment. Neither does it mean the end of an active life. As a matter of fact, there are many examples of women who started a job or a career at this stage of their lives. At least this is so in our own society, and it reflects the long life expectancy that is part of life in a modern, industrial society. A person's hereditary make-up is likely to play a role in the way that particular person's aging processes take place. But at least as important is one's way of life, that is, the combined effect, over the years, of factors such as nutrition, physical exertion, and exposure to environmental hazards. Changes in this way of life may alter the starting point of these aging processes and the extent to which they develop, as several epidemiological studies indicate (for instance, see Mausner and Bahn, 1974). Furthermore, medical knowledge and technology can curb some of the results of aging. Examples are, cosmetic surgery, surgical intervention such as coronary bypass, and the preservation or replacement of decaying teeth.

Certain chronic diseases do not manifest themselves until old age. But we now know that conditions such as hypertension and atherosclerosis (hardening of the arteries), once considered as the inevitable consequences and sure signs of aging, can, in effect, start much earlier in life. Autopsies on young American soldiers during the Korean War showed that 77 percent had atherosclerotic lesions in their coronary arteries (Mausner and Bahn, 1974: 166). It has been suggested that changes such as these should be considered either as the end result of prolonged exposure to "normal" factors in, for instance, the diet or the environment, or of a much shorter exposure to such factors in excessively high doses (see Eckholm, 1977, chapter 2, on this). If this hypothesis is true, and there are many indications that suggest this, prevention should become the first priority in our fight against chronic disease. Chronic disease is now seen as the end result of many conditions and predisposing factors that developed much earlier in life. It seems realistic to expect that, by preventing these conditions from becoming established we can prevent some, if not most, chronic diseases from developing. It is not clear whether we can arrest a process already started, let alone reverse the damage that has already occurred. This modern concept of the nature of chronic disease has led some to view old age itself as a form of chronic disease with pathological conditions developing earlier in life and finally leading to deterioration across-the-board. Can we, in other words, prevent, arrest or "cure" old age? We may indeed be able to slow its progress. However, it seems unrealistic to expect that all signs of aging could be treated or prevented by using, for instance, dietary supplements or by attempting to restore the body's hormonal balance. Further research is obviously needed before we know what causes physiological aging and what it consists of. Meanwhile, enough is already known, at this time, to allow us to explore seriously ways in which the effects of "normal" aging could be changed for the better, in its physiological, mental and social aspects. High on the list of priorities are decent living conditions, good health care, an opportunity to develop new interests (or to explore longstanding ones in greater depth) and, specifically, an opportunity to remain an active participant in one's community. Of course there is not just one way in which all this can be accomplished. The elderly need help, on the part of the Legislature, of other government bodies, national, state and local. There should be people and organizations who can take up the case for the elderly and who can show them the many resources available to them. But, most of all, the elderly are helped when they learn to help themselves, when they can set their own goals and priorities, and the more successful programs are based on this premise. The Gray Panthers are an example.

In almost every respect getting older diminishes a person's chances of fulfilling the requirements for a carefree, healthy and active life. Most people find they are slowing down and many suffer from the discomforts of "wear and tear." Many would probably welcome the chance to take it a bit more easy. But retirement abruptly ends the work routine of many years. Overnight one becomes an outsider to the accustomed circle of colleagues. Most importantly, there is usually a significant reduction in financial resources. Of course, there are positive sides. There is freedom from the more burdensome aspects of making a living. There is time to start doing the things one always had to put off till later. But, more often than not, life turns out to be rather empty without a daily goal, unless the older person has developed earlier in life, what Gene Weltfish once described as "some lifelong absorbing interest and not just a set of hobbies" (1976). Late in life the means and the energy may well no longer be there to develop such a satisfying and absorbing use for one's time. Fortunately, there are opportunities, nowadays, for the older person to find help in exploring new interests, as well as to learn about the special problems older people face, and how to deal with these. A good example is the program developed several years ago by Hope Bagger of the Gray Panthers, (Bagger, 1975 or 1976). Ms. Bagger, herself close to 80 years old, at the time, had several goals for this program. In the first place, it offered older people the chance to "update" themselves with regard to historical developments and ideas that had become a part of our society's ways of thinking during the years since these elderly students had finished their schooling. Most of these people had been too busy making a living and raising a family to keep up with new developments. Furthermore, the older people learned about the physiological and social aspects of normal aging. They became better informed about ways to safeguard their health, for instance by getting exercise and eating well-balanced meals. The program also provided political information and taught the people to become activists on their own behalf. Finally, there was an opportunity to explore new interests and develop new skills so that these people could stay alert and satisfyingly occupied during their remaining years. The older students had the option to follow those parts of the curriculum that filled their needs, which, of course, were not necessarily the same for all of them.

Programs like this can undoubtedly be found in many different settings. Not all of them will be as responsive to older people's needs as the one described. High schools and colleges have long since discovered the advantages of offering "continuing education" courses of interest to the elderly. One problem is that many of these courses are designed to help people pass their time in pleasant ways, and the elderly are not considered as serious students. Gene Weltfish, herself a most dedicated teacher until her death at 79, repeatedly warned that entertainment was necessary but that it should never be confused with the pursuit of a serious interest. It was the latter that she thought old people needed most. The ability to take advantage of these special programs depends on several conditions. In the first place, one must know about such possibilities and seek them out. This may take more initiative than some old people can muster. Then, the particular approach used, namely continuing education, must appeal, or at least be acceptable to the particular person. Also, the older man or woman must realize that, once he or she has been "updated," this process has to be kept up for it to have any lasting value. Finally, and perhaps most importantly, there should not be too many conditions that, together or separately, have to take precedence. The most important of these is the need for the necessities of life: food, shelter and clothing. Physical or mental handicaps may also form insurmountable obstacles. The lack of cheap and dependable transportation, whether public or privately organized, can put otherwise excellent programs effectively out of reach. Each locality will have its own problems when it comes to making its resources available to its older citizens. We should keep in mind, however, that people with serious handicaps, and among them old people, often are not only capable of functioning in our society, but that many insist on their right to do so. Goffman, in his *Stigma* (1963), has written most eloquently on this.

Getting older means more than all this to most women. Their specific situation derives from two circumstances. In the first place, there is their social status as women during the preceding years of life and still in force in their old age. A lifetime of having been subjected to discrimination on the basis of one's sex is bound to leave its mark. In the second place, there is the specific circumstance that these women's physiological role as the bearer of children has come to a definite end. To many men and women of the present generation over 60 this means that older women are not "real" women any more, and this again reduces their social status. During this period in their lives these women's traditional role as the person who takes care of the family is likely to end as well, and again her social status diminishes. Such a woman is no longer the mother (and the wife) she used to be. Young and Willmott (1962) have shown that, provided families continue to live in close proximity over several generations, there are traditional roles for these older women to assume. The "Mum" of such an extended family, in a working class borough of East London, remains in close contact with her married daughters and continues to help in the care of the grandchildren. At the same time the older parents' home remains the center for many of the activities that are important in a multiple-generation family, such as weddings, holidays and just routine Sunday visiting. However, when a daughter and her own family move away, the relationship with her mother changes, if only because transportation becomes too inconvenient and expensive to maintain the almost daily contact of the past. In a country such as the United States there are comparatively few multiple-generation families living close enough to make such a progression of roles possible for older women. For most women with children the end of her "active" motherhood with regard to her own children forces her to focus her energies and emotional involvement elsewhere. Her own family and her role in it have changed drastically. The so called nuclear family, father, mother and children, is considered the norm. That is the "family" people talk about when they lament the decline of the family. It was not always the norm of the American family; the increasing social and geographical mobility of the last few decades have made it easier for family members to move away from each other. Even today, this nuclear family is by no means as universal as some people and government institutions would have us believe.

The nature of the family and the roles of its different members have become a matter of great interest, specifically for women (and men) who have begun to question their traditional roles in society. We have been asking ourselves, for instance, what the essential function of a mother really is. Should we not consider this to be the caring for and raising of a child—and not necessarily by herself alone—rather than giving birth to that child and nursing it? By the same token, should we not speak of parenting rather than mothering and acknowledge that traditional activities and attitudes do not have to be assigned to just the father or the mother, but could be fulfilled by either, as the situation demands? In many contemporary families these changing concepts are being worked out. New ways of raising a family are becoming acceptable and are being shared with others looking for suitable models and solutions. However, for the present generation of older women the more traditional division of labor between men and women, in the family and in society, have been the ones they always have had to contend with. It is true, of course, that some couples managed to work out a different solution for themselves. And some women, belatedly, developed interests and careers after the children grew older. But, with or without children at home, most of these women kept performing many of their traditional responsibilities, in addition to their outside activities. Even if the family could afford domestic help, this group of working women cooked, shopped, prepared for the holidays as part of their female functions in life.

The present generation of young women does not have to accept many of these stereotypical responsibilities as somehow preordained. Many, it turns out, still do accept the more traditional female role. Some choose to do so and discover that there are many enjoyable aspects to the full-time care of and concentration on husband, home and children. But, at least, the issue has been stated and choices can be made. For many young women an old dilemma has gained new urgency and importance. How can they develop their careers and have a family as well? Among the most difficult problems to be faced is that the building of a family, as well as of a career, must necessarily be limited to a comparatively short period in a woman's life and that neither can be postponed indefinitely. The solutions to this dilemma have taken many forms. Some women decide to have no children at all. Others reject the idea that only a heterosexual relationship can form the basis for a family grouping. Some try to do what their mothers did, and work intermittently, while having their family, with their husband's consent but without much cooperation. To have a family, a career and a marriage is a very difficult undertaking, even if husband and wife share parental and household responsibilities. This may well be the hardest choice of all. It requires an extraordinary amount of flexibility, mutual understanding and the ability to "shift gears," to be able to decide on the priorities at any given time. Both spouses, and the children as well, must share in such an undertaking, and success is by no means guaranteed, even then. Some families seek the solution in communal living with other young families. In a few instances members of the extended family: an elderly parent or aunt, may be counted on to help out. But usually the couple has to depend on outside (hired) help for a large part of the child-rearing process. Of course, this last situation was always the accepted one in affluent circles. Among the poor there never was much choice: having the mother stay at home, taking care of the house, husband and family, was a luxury they could not afford. In other words, among the poor, the division of labor between the sexes, to which a large part of our society still pays lip-service, was superseded by the division of labor reflecting economic status. We can say that poor women became available to take over some of the traditional female roles in other people's families, and the traditional sex roles did not change at all. Leith Mullings, discussing the causes of "gender asymmetry" in another context, wrote: "In a society where profit is the driving force, anatomical differences between males and females, like culture history differences between ethnic groups and phenotypic differences between populations, are utilized in such a way as to support that system and maximize profits. Where such differences converge, as is the case with Afro-American women, oppressions will be particularly intense" (1980: 26, 27).

This discriminatory division of labor between males and females determines the social status of women, young or old, in the home and in the work force. It also determines the financial resources a woman can call her own. This is so as a member of the work force. It is even more so as a housewife or as the widow of a working man, where she will only have rights to a part of her husband's wages. Chances are that for a working woman both social status and earning power will be less than for men in a comparable position. Consequently, the basis for her pension will be lower, so that this particular distinction between men and women becomes particularly important when the woman retires. In short, whether she worked or not, the reduction in living standards, with its consequences in available funds for shelter, food and health care, tends to be even greater for old women than for old men, specifically if they are single.

A further consideration presents itself: If the division of labor between men and women is based on gender (here defined as the social attributes of one's physiological sex) it should reflect their differences in social status. However, the obvious physiological and anatomical sex differences (the genetic makeup) have always been used as a basis for justification of this difference in social status. Social discrimination on the basis of sex consequently acquires the character of being inevitable and unalterable. If a woman's social position is measured by the yardstick of her "traditional" role in the home, this implies that her true value lies in work that cannot be compared to "real" work, that is with work for wages in the market place. Therefore, any work she does do outside the home is rated on the basis of her "non-work" at home, useful as that may be for her immediate family. From the start a double standard exists for comparing men's and women's work. Being a housewife of many years certainly does not count as work experience when she applies for a job. Neither does it help her when she tries to get into graduate school to continue an education interrupted by years of care for the family. Continuing education is available to these women, as a matter of fact they are part of an eagerly courted group of potential paying students. But the courses that are offered are, more often than not, what the particular institution considers suitable, rather than the challenging and demanding subject matter that would help prepare her for that new start she is looking for.

Because a housewife did no real work she is entitled neither to a recognition as worker nor to full compensation for the work she did. Her pension-as-housewife consists of the right to share in her husband's pension. As the widow of a federal employee, for instance, she would receive about half, while he would keep the full pension regardless of his marital status (See Pension Regulations for Federal Employees, U.S., July 1975) Private pension plans vary in detail, but the principle remains the same: The non-working spouse loses out when the one whose work earned the pension dies. The case for Social Security is somewhat different. The spouse of a person getting Social Security income is entitled to a check in their own right, provided they have reached the legal age. After divorce or remarriage these checks for dependents will be stopped, with the following exceptions. A woman of 62 or older will not lose her right to this check after divorce, if she was married to the worker for 10 years or more. A widow or widower over 60, receiving dependents' benefits will not lose them when they remarry. These were the regulations prevailing in 1979 (HEW, January, 1979). At the time of this writing the whole system of Social Security benefits and rights is being challenged as overly generous and as a sure prescription for bankruptcy. Critics of this position have pointed out that situation derives from the fact that the United States' Social Security system is expected to pay its own way. This is in contrast with the situation in most Western European countries, where payroll taxes used to underwrite comparable programs are supplemented from general tax funds. Warren Weaver Jr., writing in the January 17, 1982, *New York Sunday Times,* warned that we should not consider these countries extravagant in their care for the elderly, as compared to a much more prudent United States. The truth of the matter is that the United States is only able to follow this course of action because its population over 65 is much smaller by percentage points than that of the European countries. Not until early in the next century will our population of elderly reach comparable figures (p. A 21).

Men can become victims of the inequities in the system as well. Because women were rarely acknowledged as being potential breadwinners, surviving husbands who depended on their wive's earnings had even fewer rights than housewives depending on their husbands. Government regulations and government pensions, including Social Security benefits, no longer distinguish between the sex of the wage-earning spouse and that of the dependent or survivor. However, for the present generation of older women their wage-earning potential was never equivalent to that of the male workers and their pensions reflect this difference.

If housewives did no real work, it follows that they can not really retire. Indeed, I have heard of cases where the wife of a retired businessman was driven to distraction by his insistence on standards of efficiency in the home that he had always demanded from his secretary in the office. Neither does a retirement age seem to apply to housework when it is being paid for. In contrast to members of more prestigious professions a housewife, cleaning woman, nurse maid or housekeeper is allowed to work until she is no longer physically able to do so. In this respect their work resembles certain types of unskilled labor performed by men, and also that lowliest category of work our money-oriented society knows, that of the volunteer worker. Our society can not exist without the unpaid or underpaid labor contributed by all these people, but there is little recognition or compensation for these services.

The tendency to equate a woman's social status with her powers of reproduction, and as such being a direct function of her chronological age, leads to yet another inequity. The age at which a man can father a child, or convince a young woman to live with him, does not show a definite cut-off point. But not until recently has the fact become accepted that older women can—and do—enjoy sexual relations, if given the chance. The ability to be a satisfactory sex partner has nothing to do with a woman's ability to conceive. Society appears now willing to accept marriage between old people. It may even concede that elderly marriage partners can have enjoyable sexual relations (Blythe, 1980: 259). Many people still have trouble understanding that old women have sexual needs, however. Somehow, such a need is easier to accept for old men. De Beauvoir has written most eloquently about this (1972: 347 ff). Until quite recently, most textbooks on gynecology and surgery considered the cessation of a woman's ability to reproduce the most important argument for radical hysterectomy.

The time of retirement may force older couples to re-examine their stereotypic concepts of work and pastimes as being only suitable for either men or women (Kalish, 1969). There may be a belated chance for such a couple to become more flexible and, as Butler (1975) has suggested, to grow, either together or separately, as people in a further stage of their life cycle. The adjustments the retiring husband must make may well be greater than for his wife. To some extent he now becomes a part of his wife's long-established domain. Every elderly couple will have to work this new situation out in their own way. The first requirement is that they recognize the need for this and are willing to work at it. Those solutions seem to be most successful where both partners are able to give up some of their old roles without feeling this as a threat to their identity. I knew a couple where the wife expected the husband to lend her a hand and do the vacuum cleaning. He, on the other hand, expected to be waited on, hand and foot, as during the days that he earned a good living in the construction business. They were both quite dissatisfied with each other. Finally, the wife convinced the husband to move to Florida, where their daughter lived and where the more casual life style may well have made it easier to adjust to retired life.

Single women, when they get older, could well have the advantage over their male counterparts and over married women. Provided they have sufficient financial means and are in reasonable physical and mental health, these women may enter upon a period of independence from other people's demands and start pursuing their own interests in a determined way. In the first place, much more so than the others, they will have a long experience in creating their own physical and social environment. Secondly, being more inclined than men to seek medical advice (Mausner and Bahn, 1974, 48), they may reap more benefits from the development of geriatric medicine. At least, if these women are willing and able to deal with the traditional sexist attitudes of most medical care providers. For some of these women professional prejudices based on sex, age and class or "race" may make their experiences in doctor's offices or hospitals very painful indeed (Ehrenreich and Ehrenreich, 1970: 13–15). Finally, older women may benefit from the growing public acceptance of single women, young and old, dressing and going as and where they see fit.

The situation just described only fits the ideal case. Many single older women do not have sufficient financial means, and should their health fail they may not be able to get the help they need. With the federal and state governments cutting down on Medicare and Medicaid, the health provisions for those on Social Security and for the indigent, respectively, may well fall short of the minimally necessary. Poverty is the most serious obstacle to a satisfactory old age. As we have seen, elderly women, even more so than elderly men, frequently lack the funds needed to lead a life based on the bare necessities. The life expectancy for members of our society is increasing, but appreciably more so for women than for men. In the period from 1900 till 1971 the life expectancy for white males increased from 48 years to 68.3 years. White women's life expectancy rose to 75.6 over that period. By comparison, non-white males' life expectancy was only 61.2 in 1971. (Mausner and Bahn, 1974: 199). As women can be expected to outlive their male contemporaries, the number of single women among the elderly is growing disproportionally. This fact, added to their generally lower socio-economic status, makes this particular group most vulnerable to inflation and to cutbacks in social services on the federal and state levels. The proposed abolition of the minimum benefit under Social Security would hit this group of elderly particularly hard.

The final part of this paper will address itself to some of the health problems of older women, in which they are known to be (or are presumed to be) different from men. It may be possible to pinpoint some ways in which these differences can be related to gender rather than the genes. That is, we may be able to consider with some justification whether it is not their life-long treatment as women at the hands of society rather than the anatomy and physiology of their sex, that is at the basis of these differences. I will discuss briefly two instances: hypertension and its complications (heart disease and stroke), and lung cancer. For both diseases the morbidity and mortality rates are considerably higher for men than for women. As a matter of fact, it has been customary to explain these differences in more or less "sexist" terms: either the women are considered to be protected by their female hormones, or they are presumed to have led a much less stressful life than men. Explanations such as these may well be far too simplistic. Even more simplistic is the custom to blame "racial," that is hereditary, factors when further differences are found between "racial" or ethnic groups within each sex category.

These "racist" and "sexist" explanations appear to be telling examples of the lack of sophistication of the epidemiological imagination. For *hypertension* the following is known (see also Vroman, 1980, 193–196). In the United States twice as many Blacks develop hypertension than Whites, and this holds true for all age groups and for men as well as women. Young Black men tend to develop hypertension at an earlier age and in a more severe form than any other group in our population. Fewer women die of hypertension and its complications than men, even if their blood pressure values are similar (Freis, 1973: 61–68). Lew (1973, 46) writes that Black women have a high incidence of hypertension under the age of 45, as compared to White women. After that age the incidence for White women increases while that for Black women remains the same. Lew attempts to explain the differences between Blacks and Whites on hereditary grounds. He cites a study where Blacks were matched by occupation and income to a comparable group of Whites. It was found that, regardless of similarities in socio-economic conditions, many more Black subjects than Whites developed hypertension and died from its complications. However, no mention was made of the possibility that life-long dietary habits (such as a high salt intake from one's childhood years and obesity as result of high caloric intake) could have played a role. Neither was the possibility considered that even successful Blacks may have had to cope with considerably more stress for many years than their White counterparts to attain a comparable socio-economic status. One of the most devastating complications of hypertension is a stroke, or cerebrovascular accident (CVA). This is the third most common cause of death in our country, but over the last 50 years a steady downward trend has been observed. Levy (1979: 490) writes that this decline in mortality has been found for "all four major age-sex-color groups . . . but the decline by sex for non-whites (primarily blacks) is relatively and absolutely greater than that for the comparable white population," and this is especially so for non-white women between 35 and 45 years of age. The reasons for this phenomenon are not clear, but may include earlier detection and treatment as well as a response to the recent wide-spread effort at educating the general public to the dangers of high blood pressure. In short, data such as these have to be interpreted with great care. Rather than seek explanations in genetic characteristics related to "race" and sex, it may be more prudent to look into environmental factors, including those deriving from the social environment. One of the obvious advantages of such an approach is that it can lead to attempts at prevention while the genetic hypothesis accepts the status quo.

Lung cancer has been strongly associated with cigarette smoking, although air pollution and occupational exposure to substances such as asbestos fibers and silicates have also been linked to the disease. When such exposure is combined with cigarette smoking the effects of each are greatly enhanced (Eckholm, 1977, specifically chapter 6). Mausner and Bahn (1974: 76, 77) write that death from lung cancer is by far the most common form of cancer for males in our society, and that the mortality rate has risen alarmingly over the last 30 years. For women, on the other hand, mortality for lung cancer is comparatively low and lags far behind that for breast cancer. However, there has been a small but significant increase in deaths from lung cancer in females in the last 15 years, a trend which may be attributable to an increase in cigarette smoking among women (*ibid*: 100, ff.). The conclusion of these very careful epidemiologists is that there is a great deal of evidence that points to cigarette smoking as the factor mainly responsible for the steady increase in male mortality from lung cancer, and for the change upward for that in women. There are, however, still some investigators who consider the connection between cigarette smoking and that of lung cancer as circumstantial, and maintain that the difference between the sexes is due to "constitutional" factors. In other words these people relate the overall low lung cancer mortality in women not to their smoking habits but to certain protective influences deriving from their (biological) sex. Here, again, the hereditarian hypothesis would accept the *status quo*, while the environmental hypothesis would search for specific causes and would advocate preventive measures.

In summary we can say that the respective roles of genes and gender have an influence on women's health throughout their lifetimes and express themselves in characteristic ways in old age. As yet we are not able to distinguish clearly between these two influences, and this may well remain the case for a long time to come. One possible contribution may come from careful comparisons between the present generation of older women and those of the future. In time we can hope to learn more about just how gender roles and expectations have influenced the way we conceive of the nature of hereditary factors and how these influence our lives as women. In the meantime it would be the more prudent way of looking at women's health if we investigate as many as possible of the many circumstances that determine it. Although we cannot deny that women are female, in the genetic and in the social sense, this is not the only determinant of their health status, and it may well be a form of sexism to stress this circumstance above all others. Both men and women will be better served if we remain alert to the many environmental factors that influence our health.

REFERENCES

Bagger, Hope Sabin
   1975   A study program for the elderly and friends of the elderly. Philadelphia, Gray Panthers.

Blythe, Ronald
   1979   *The View in Winter. Reflections on Old Age*. New York, Penguin Books.

Butler, Robert N.
   1975   *Why Survive? Being Old in America*. New York, Harper and Row.

Butler, Robert N. and Lewis, Myrna I.
   1974   Life-review therapy. Putting memories to work in individual and group therapy. Geriatrics; special issue: Symposium on mental health and aging: life cycle perspectives. November, pp. 165–177.

De Beauvoir, Simone
   1972   *Old Age*. London, Andre Deutsch, Ltd and George Weidenfeld and Nicolson, Ltd.

Dubos, Rene
   1965   *Man Adapting*. New Haven, Yale University Press.

Eckholm, Erik
   1977   *The Picture of Health*. New York, Norton.

Ehrenreich, John and Barbara
   1970   *The American Health Empire. Power, Profits and Politics*. New York, Vintage.

Freis, Edward D.
   1973   Age, race, sex and other indices of risk in hypertension. Laragh, John H., Brunner, Hans R., Sealy, Jean E. (eds) *Hypertension Manual*. New York, Dun-Donnelly Publ. pp. 31–41.

Goffman, Erving
   1963   *Stigma. Notes on the Management of Spoiled Identity*. Englewood Cliffs, N.J., Prentice-Hall.

Health, Education and Welfare, U.S. Department of Social Security Administration.
   1980 Jan.   *Your Medicare Handbook*. SSA Public. No. 05-10050.

Kalish, Richard A. (ed)
   1969   *The Dependencies of Old People*. Institute of Gerontology Series No. 6. Wayne State University and University of Michigan.

Levy, Robert I.
   1975   Stroke decline: implications and prospects. The New England Journal of Medicine 300: 490, 491.

Lew, Edward A.
   1973   High blood pressure, other risk factors and longevity: the insurance viewpoint. Laragh, John H. et al (eds) *Hypertension Manual.* New York, Dun-Donnelly Publ. pp. 43–70.

Mausner, Judith S. and Bahn, Anita K.
   1974   *Epidemiology.* Philadelphia, W. B. Saunders Co.

Mullings, Leith
   1980   Notes on women, work, and society. Tobach, Ethel and Rosoff, Betty (eds) *Genes and Gender III.* New York, Gordian Press pp. 15–29.

U.S. Civil Service Commission. Bureau of Retirement, Insurance, and Occupational Health. *Your Retirement System.* Pamphlet 18.
   1975 July

Vroman, Georgine M.S.
   1980   *The Anthropological Dimension in the Rehabilitation of Aphasics.* Doctoral dissertation, Graduate Faculty, New School for Social Research, New York.

Weltfish, Gene
   1976   Course notes, Anthropology of Aging. Graduate Faculty, New School for Social Research, New York.

Young, Michael and Willmott, Peter
   1962   *Family and Kinship in East London.* Pelican Books. Harmondsworth, Middlesex, England, Penguin Books.

SEXISM AND RACISM IN HEALTH POLICY

June Jackson Christmas, M.D.

The title of my paper refers to those who are affected by the combined and negative effects of sexism and racism. Whereas much of what I say will be applicable to many women, I wish particularly to emphasize the situation that applies to ethnic minorities of color who are women. If our situation is bad now, it is becoming worse, not only worldwide, but in this country. It is worsening because of the kinds of health policies that exist in the United States.

Women occupy a devalued imposed social position in most societies: this is certainly true of our supposedly egalitarian society. It is a devalued position that restricts our roles as women—all of us, black, white, brown—and that has a negative impact upon our health status and our access to appropriate services, on our educational and employment opportunities as well as on our economic status. If these limitations are true for all women, they are even more true for ethnic minorities of color. By law and/or custom, depending on where we live, we are in positions of jeopardy. So, that is the first point: the devalued, imposed social position that we occupy as women in society is worse, if one happens to be a minority woman.

The second fact that I use as a premise is that women are poorer than men. We know that women are in positions of less power and control than men. We know also that we earn less, even for comparable work, and have less access to the goods, services and power that money can buy, including health care. These factors are even more important for ethnic minority women of color. We know that this second burden is borne by many women of color, for they are victims of sexism and they suffer also from greater poverty relative to men in their society.

In addition there is a third factor, that is, the significant factor of racism and oppression which is inherent in our Western society. Dr. W. E. B. DuBois wrote long ago of the critical problem of the twentieth century—the issue of color and race, the problem of racism. The problem exists; it steadily worsens. We see it not only in the South, but here in New York City and across the United States. We see it in those countries which the United States supports, such as South Africa, in apartheid. We see it in the tragedies of the oppression of colonialism in Latin America. We see it in the fascist discrimination of blacks against blacks in Haiti. We know that this racism places people of color disproportionately among the ranks of the poor and downtrodden. And, of course, in the United States, we who happen to be black and women have a history in which we have been objects: we have been used as breeders of cannon fodder and as breeders of people to labor without compensation in the fields. We have been used by white society and we suffer from the remnants of these experiences.

Black women were viewed and used as the epitome of the embodiment of forbidden sex. White women also lost something in their status on the pedestal of purity during slavery days. White women also suffer because they were denied their own sexuality. If we are Native American women, we have been exterminated and have seen our children snatched from us to be "educated" away from us, someplace else, in the reservation school. Until recently, we who are Caribbean women experienced the discrimination that allowed us to come in illegally to labor, or to be enticed in to do the dirty work, but which kept us from becoming citizens. And, of course, as black women we know that today, whether we live in Atlanta or in Brooklyn, our children are at hazard and at risk, of disease, deprivation or death. What happens in Atlanta may be the most vivid violence, the murder of children, but we must not forget that there are scores of black and brown young people harassed or killed by police each year in this country; there is no fair investigation of many of these acts of violence. We still experience racism and its violent results.

In addition, in the United States a patriarchal society and resultant laws provide the foundation for oppression that is economic and social, racist and sexist. Oppression pervades the total society; health risks and hazards result in all aspects of our lives. I shall refer to the work place only to say that all of the damaging experiences and exposures to risks which were highlighted in earlier presentations are worse for those people, both men and women, who labor at the most menial jobs, who *have* to take the dirtiest jobs; they are frequently minorities and other poor people.

We see these inequities in the justice or "injustice" system that we call our correctional system. Few facilities exist providing adequate health and mental health care for inmates; this is especially true where women prisoners are incarcerated. Brutality is the general experience; black and brown women and men are disproportionately represented in prisons. Human service workers pay scant attention to the fact that when people are kept prisoners they both experience and impose brutality.

In so many aspects of our lives sexism and racism exist; they exist in an economic situation in which policies of the big business of the health industry produce severe problems for black and brown people, for women of all races, but most dramatically for poor, minority women. Women head households in increasing numbers in all races and ethnic groups. Today, of the families headed by women, 70% are white, 28% are black, 6% are Hispanic in origin. If we consider their income, we note that black families headed by women are at the bottom of the economic barrel. Now, health policy in this country requires that money buy health care, although not always the best qualitatively. Indeed, money is needed for access to health services, unless one is doomed to receive health care that is devalued and that may also be poor in quality. Health care is not considered a right. Health insurance is neither universal nor comprehensive. That is a terrible reality. It does not *have* to be, but it *is*. But the press for a national health plan has almost been abandoned. Instead, people are told to rely on private insurance obtained often through the "generosity" of employers, or they must "spend down" until they are eligible for Medicaid, which itself becomes more restrictive day by day. The working poor are caught in a bind.

Economic deprivation and societal pressures interact to reinforce sexist and racial values. Look at the issue of black women and work: 52% of all black women work, *not* all by choice. Yet, as Afro-American women often say, we were the first to be involved in the act of women's liberation. We have always accepted the idea that we would work. I never assumed that there was any different expectation. I grew up in a middle-class family, but I always assumed that I would work. Nevertheless, the median income of all black women is only $6,600. That is 94% of the income of white women; it is 73% of the income of black men and 54% of the income of white men. So the basic problem of limited access and maldistribution of health care, and the uneven quality of care affects even more dramatically people who are poor, and who happen to be of minority background. It affects women even more if they are heads of households with hopes for work outside of the home, but with few opportunities for adequate child care that might allow them to join the labor market.

There are those who would state that money will nevertheless buy good health care and, therefore, ensure good health. However, in spite of the fact that the middleclass and wealthy in the United States can buy relatively good health care, the health status of our country is still in need of improvement. We lag behind the Scandinavian countries in several indicators. And we are losing ground. If you look at infant mortality, for example, the United States has slipped from fifth place overall in 1950 to a point where, in 1976, fourteen other nations were ahead of us. For minorities those statistics are even worse. Infant mortality among blacks in this country is highest of all, exceeding even the rate among American Indians (who finally received priority and long overdue Federal attention, funds and programs). The rate of infant mortality in the South Bronx, where some of us in this room work, is higher than that in the developing country of Cuba.

I received my medical training in Boston and in New York City. There I saw the effects of racism and sexism in medical practice. Racial minorities, often being poor, are effectively segregated on wards of underfunded public hospitals. Poor children are four times more likely to have chronic medical conditions as those who are not poor. Very few minority physicians are allowed to be created by our medical education system. If we think that practices emerging from policies which have made piecemeal provisions so that people can purchase health care have remedied inequity, we are wrong. Medicaid is supposed to be distributed equitably. Yet, in the South in 1976, the average Medicaid expenditure for a white person, male or female, was $350.00; for a black person, the expenditure was $250.00. Medicare, too, is utilized unevenly. So benefits of the little patchwork insurance system we do have go more to whites than to minorities. Lack of knowledge of eligibility, restrictions imposed by racism, and inability to maneuver the bureaucracy are all at fault.

And health status suffers from the twin burdens of racism and sexism. Looking at life expectancy, a white *middleclass* woman, born in New York City, can expect at birth seven more years of life than a black woman of the *same* socio-economic class born in the same city on the same day. Robbed of seven years of one's life and then exposed to the differential in employment opportunities, education, housing and access to health services! We know that these facts are part and parcel of the United States health policy which is based on power, and which is in the hands of men who are predominantly white. Health policy is dictated by large corporations in the business of the health care industry, by drug companies in the business of health care and health services, by hospitals and by our professional guilds, where decision-making and the important process of influencing policy is denied many people. Health policy emphasizes the sickness-model, curative and treatment-oriented, and minimizes public education and health education. The philosophy is based upon negatives, upon illness. We do not have a philosophy that speaks toward health, toward discovering the strengths within people. We have a philosophy based on illness, on the treatment-of-sickness service system/nonsystem. It mouths platitudes about prevention and, yet, does not provide financing or reimbursement for preventive programs. Health policy neglects rehabilitation, which views the whole person in a social context; after all, rehabilitation helps people live in a society. Our idea of health care is based on a narrow medical model that is individualistic rather than social, sickness-, rather than health-, oriented.

So, on the firm base of racism this translates into programs which do not deal with the cultural background of people. We see this neglect at every turn. I recall that when we in the New York City Department of Mental Health, Mental Retardation and Alcoholism services started programs for alcoholics in Hispanic neighborhoods we urged that agencies hire counselors who were Spanish-speaking and who would incorporate their own background into their rehabilitative activities. The

State Mental Hygiene Department said: "Oh, you can't just go looking for Spanish-speaking workers." When we spoke about programs we had established in the black community and stated that we needed to give them more technical assistance because racist practices had denied them opportunity to acquire these managerial skills, we were also told: "What? Give extra services to one group or another?" Fortunately, we were able to persist and establish such procedures as policy. Culturally relevant and ethnically appropriate services need to be provided. Yet, it is distressing that the Carter Administration failed to go along with such recommendations to address inequities, recommendations made by several commissions on which I worked, including a Task Panel of the President's Commission on Mental Health. With the Proposition Thirteen mentality, with the rejection of social concerns, and with the lack of commitment to minority rights and women's rights of the Reagan Administration, the situation will become worse.

What might happen? Let's look at some of the Federal policies, because they are most dramatic. Sterilization abuse will likely mount. Respect for patients' rights is under attack. I attended medical school in the days when surgical experience was expected to be gained on the bodies of black and brown people. Unnecessary operations were conducted by uncaring men upon women, by uncaring white people on blacks. These resulted in many young black women having hysterectomies and other procedures, when they were still in their twenties; surgery took place for disorders for which comparably ill white women received other kinds of treatment. And no one questioned medical abuse and experimentation. There was widespread sterilization abuse, for example, in Puerto Rico, where 33% of the young women were sterilized. These abuses may increase, including disregard of the sterilization guidelines that we fought so hard for in the last few years, of the rights of mental patients who have been treated as less than human, of the lives of persons in institutions, including prisons and juvenile detention centers. With a conservative Supreme Court and an anti-regulatory mood, safeguards will no longer be considered of Federal or even of public concern. It is true that we need to watch the watchdogs, but we have no information to suggest that states rights and the private sector will guarantee the health concerns of those who have been victims of discrimination. On the contrary, their philosophy has favored those in power.

The Federal Drug Administration, which is supposedly the repository for the public good, allows drug companies to conduct studies on efficacy of drugs such as DES, because the drug companies have power. Many of these studies are carried on by physicians and others still using poor black and brown women, using minority people, as human guinea pigs. Experimental safeguards may be further weakened. Of course, this philosophy is carried out through the rest of the world also. For example, although the FDA says that DES and other drugs cannot be used in this country any longer, they may still be marketed in the Third World, in Africa, Asia, and Latin America. When a multi-national conglomerate says it does not market its infant formula here as it does in Africa, that is true. I know what they do in many African countries. Last year I saw little babies dying of malnutrition in a Kenyan hospital. Their mothers had been persuaded by Kenyan women dressed in nurses' uniforms (trying to make a living) selling infant formula to women who did not have clean water and sanitary facilities. So these babies were dying.

We all know the furor concerning tampons and toxic shock syndrome. Tampons, used in our bodies, are still tested only by the companies themselves, and not by the FDA, because the FDA has not declared tampons to be a device to which considerable risks are attached. You know very well that if they were used in men' bodies, they would have been tested differently. Even the limited regulation that we have in the Food and Drug Administration is going to be destroyed and damaged in the anti-regulatory mood that is now the hallmark of the Reagan Administration.

When we come to an issue such as the access to health care, we know that minority women are in a worse situation. We know also that their ability to choose abortion or to have a child will depend, in part, on their hopes for the kind of life that child will have to live and for the opportunities for growth and development which it will have. If they live in a city such as New York, what are their chances? Where needed health facilities are being closed in minority neighborhoods, where the political leadership, all majority race, follow the example of the Mayor in acts and statements which are insensitive, obstinate and bigoted, what hopes do they have? We live in a city where health policies are made by people who care not for interests of the human beings of the city, white and black, but instead, for the banking, real estate and financing interests; in the majority themselves, they attend only to the majority in power. And the effects of their policies are increased dehumanization of health and human services. We live in a city where the head of the Health and Hospitals Corporation makes an anti-black, racist slur at a meeting and his biased remark is excused by the mayor. You know and I know very well that, had it been a black official who made, for example, an anti-Semitic statement, he would have been out of that job (on the invitation of the same mayor who was so quick to excuse his prejudice) before you could cry racism.

Indeed, we live in a city and in a country where increasing racial polarization is tolerated and fostered by leadership which is in the service of the wealthy and the white. While ignoring the root causes of social unrest felt by the poor, and angrily expressed by alienated minorities, they call for more law and order, repression and cutbacks. Policies that already lead to family breakdown influence young black people who are disenchanted and frustrated to work out their anger on themselves through mounting suicide and alcoholism. This is expressed by young blacks through anger and violence directed not only outwardly but increasingly toward their in-group. If we look, we see a hopeless future for generations of black children of black women, children who are uneducated or poorly educated; that future becomes more desperate. It worsens not only for black women and their families, but for the communities in which they live and for others in the wider society who are affected by these social distresses. But these conditions receive scant attention by way of prevention or of remedy. To control the social epidemic or disorder may appeal to those seeking quick solutions. It will not make up for the history of deprivation and racism spawned by slavery and nourished by present day remnants in Jim Crow, whether in health, housing, politics or the economy.

The uncontrolled illicit drug industry has already killed thousands of minorities or so damaged their lives that they are effectively lost. This drug industry that exists and flourishes on 116th street in New York City is not seen by the police;

(substitute your own street or neighborhood where business is carried on). It is a drug industry that is making millions of dollars for the unseen drug peddlers downtown. Of those young men who returned from Vietnam, many of whom are the sons of black and brown women, a high percentage came back addicted to drugs, unemployed and disenchanted. And the reachout centers staffed by Vietnam veterans are high on the list of programs to be eliminated by Reagan, along with alcoholism, drug abuse and mental health services, including the very behavioral science research which might contribute toward some solutions for these tragic social problems.

The discrimination that affects the wide society has an even worse affect on those who are young. They grew up not having experienced that excitement of the Sixties and of the Civil Rights Movement, not having seen the occasions when people began to take charge of their own lives and have some input into health policies. The Model Cities programs were done in by bureaucrats, by politicians, and by lack of funding. The input that people might hope to have through that local planning group of the Health Systems Agencies already as a matter of policy has been weakened by the last Congress which said: "You no longer have to reflect proportionately the peoples of your community. You don't have to have poor people or Hispanic or black people in any relationship to their proportion in your population. You just have to have some there." The trend will continue, away from responsibility for the public, accountability to the public, and involvement of the public.

If I am gloomy, it is because I see in our recent history fleeting attention to one important problem replaced by passing involvement with another as though social issues were fads. The Civil Rights Movement, which gave the Women's Movement some of its techniques, was soon to be replaced by the fight against an unjust war, and, indeed, it was a terrible, unjust war. Then along came some passing concern with ecology and then concern with women's issues, and now concern with energy. All of these things are important to our health, to our wellbeing, to our existence as a nation. They must all be seen as vital and interrelated. To view them otherwise could be a tragedy. We are concerned about what happens to women; to their increasing alcoholism, which is higher in black and brown communities; to spouse abuse and rape (and most of the people who are raped in this country are black women); to all the violence that exists as one essential element of racism.

If we let our concern with all of these issues and problems that affect women as a whole make us think that women's issues are white issues, we are wrong. I speak often before feminist groups, and I come bearing a message that women come in all colors and the health problems that we talk about are indivisible. Unless we begin to make sure that we do not separate ourselves into issues of women and sexism as a separate from issues of race and racism as though those two did not belong together, each demanding our special attention, coalitions of interest and concern, we will find that not only will we have less effect on health policy, but we will be leading towards a fragmented people, and an unhealthier nation.

REFERENCES

Christmas, J. J.
    1978. "Alcoholism services for minorities: training issues and concerns" Alcohol and Health Research World. 2:20–27

Christmas, J. J.
    1977. "How our health system fails minorities" Civil Rights Digest. 10:2–11.

Christmas, J. J.
    1975. "Unmet needs as they relate to drug and alcohol abuse by women" presented at "Drug, alcohol and Women: A national forum" sponsored by N.I.D.A.

Christmas, J. J.
    1980. "Women, health and society: an overview of women in health care" presented at NGO Forum at the World Conference of the United Nations Decade for Women, Copenhagen, Denmark.

EPILOGUE

Betty Hughley and Betty Rosoff

In order to try to sum up what this book is all about, we must come back to the title *The Second X and Women's Health*. The second X chromosome is the only chromosome difference between women and men, that is men have an X and Y while women have two X chromosomes. The other 44 chromosomes are the same in men and women. Is this enough to explain the sexism that women suffer in the whole area of health.

Victoria Freedman in her paper discusses how genes are only the beginning of a process. Therefore the expression of the genes occurs in the environment of the cell and this cell environment is integral to gene expression. If the cell can effect what the gene does, then the internal environment of all the cells of the whole organism and on another level, the external environment in which the organism lives must play a role in the kind of organism that develops and functions. Therefore differences between different individuals must result and since each internal and external environment is different the expression of the "second X" cannot be the same for all women.

Genes help produce the sex hormones that affect a woman's physiology so that she can function reproductively. Susan Gordon emphasizes that both female and male hormones are found in both sexes and seem to have functions in both, though the levels of each vary. Many recent studies have shown the presence of these hormones in the brain; however, it is only in the hypothalamus that their function is clear and can be correlated to a behavioural pattern. There is no scientific evidence yet that the cerebral cortex, the site of gender thought and behaviour, is also affected by different levels of androgens and estrogens.

The anthropologist, Jagna Sharff, further equips us to shoot down the false premises on which our male dominated society is based. Egalitarian societies of the past, before they were exposed to colonial rule, had equality of the sexes. When we change the system that imposes sexism and racism for greater profit for a few into an egalitarian society without wages and profits, equality of the sexes will be achieved.

Ethel Tobach exposes the pseudoscientists and those who abuse science in order to maintain sexism in our society. Her analysis shows that the genetic determinists (sociobiologists) provide the false theoretical bases for the poor health care that many women and men receive in our society.

Toxic chemicals, whether they are the culprits in mass psychogenic illness (see Ben Harris) or whether they cause other health related conditions in women, fetuses or children (see Judith Bellin) must be detected in the workplace and their release stopped. The NYCOSH study guide describes how workers in a shop can do this *and can organize to fight this profit making contempt of* the people. The exposure to occupational hazards is not confined to the work place. Workers are only the group most directly and intensively exposed to its harmful substances and conditions. Members of their families are the ones next in line to the assault by these environmental dangers. Eventually the harmful waste products will invade the total environment. Anybody, either directly or indirectly through someone close and dear, can become a victim. The fight against occupational hazards concerns us all.

Though we try preventing illness, once we are sick we must contend with the sexist attitudes of the medical profession. Marji Gold suggests that the training of doctors perpetuates the sexist myth of women as hysterical whose illness is all in the mind. How can a woman get proper medical care from a doctor who thinks her sickness is imaginary? Racist and sexist prejudices affect the medical care of black women the most, but we all suffer as a result of racism and sexism. Therefore man and women, black and white must fight together for good medical care available to all. June Christmas emphasizes this point also and makes clear that the struggle for adequate public health programs must continue until the federal government assumes its proper responsibility for the health of everyone.

The approach of psychologists and psychiatrists towards the mental health of women is also influenced by genetic determinism as is evidenced in the papers by Myra Fooden and Susan Bram. Bram discusses the role of childbearing on the mental health of women and again exposes the psuedoscience that says that women suffer psychoses because they do not perform their reproductive functions. Some women who have children may suffer postpartum depression for social or economic reasons, while other women may choose to be childless and experience good mental health. Mental health for both men and women is a function of the society in which they live and their relationship to it. While the special problems of the health and medical care of older women must be addressed, Georgine Vroman points out that poverty is the biggest problem. No medical care can make a person well if there isn't enough food to eat and the pensions of former "housewives" is below poverty level.

Our present society based on the exploitation of the majority by the profit making minority does not encourage the active participation of all women. Young women, middle aged women, older women should not be restricted to the home for their mental health, their physical health and the good of society. The participation of women is needed in all phases of life and most particularly today in the struggle to change the conditions that make for poverty, war, inequality and poor health.

SOME ADDITIONAL RESOURCES FOR MATERIAL ON WOMEN'S HEALTH

Barefoot, Sara, Margaret Jee and Danielle Lesser (Eds.)
What Makes Women Sick?
Science for the People
Published by the British Society for Social Responsibility in Science
9 Poland Street
London, W1V 3 DG

Boston Women's Health Book Collective, Inc.
West Somerville, Mass.

Brodsky, Annette and Jean Holroyd, Co-chairs of Task Force
Report of the Task Force on Sex Bias and Sex-Role Stereotyping in Psychotherapeutic Practice
American Psychologist, 1975, 30, 1169–1175.

Farnes, Patricia M.D. and Eugenia Wild Schweers, B.A.
Women and Health Care: A Model for Women's Studies
Journal of the American Medical Association, 1980, 35, 182–18
Women and Health Care III: What are We Missing? by Patricia Farnes, Barbara E. Barker and Eugenia Wild Schweers; write for preprint to: Dr. Patricia Farnes, Brown University, Providence, Rhode Island

Hilberman, Elaine, M.D. and Nancy Felipe Russo, Ph.D.
Mental Health and Equal Rights
Psychiatric Opinion, August, 1978, Pp. 11–19.

Issues in Health Care of Women
Journal published by Hemisphere Publishing Corp.
1025 Vermont Ave. NW
Washington, D.C. 20005

Marsh, Jeanne C., Mary Ellen Colten and M. Belinda Tucker (Eds.)
Women's Use of Drugs and Alcohol: New Perspectives
Journal of Social Issues, 1982, 38, #2

Nadelson, Carol C. and Malkah T. Notman (Eds.)
The Woman Patient
A Series of Books published by
Plenum Publishing Corp.

National Women's Health Network
224 Seventh Street SE
Washington, D.C. 20003

Reinharz, S.; Marti Bombyk; Jan Wright (Eds)
Feminist Research Methodology in Sociology and Psychology
University of Michigan, June 1982

Sexual Harassment Kit
Federation of Organizations for Professional Women
2000 P Street, NW, Suite 403
Washington, D.C. 20036

Women in Crisis
37 Union Square West
New York, N.Y. 10003

Women's Issues (in Mental Health)
American Journal of Psychiatry
1981, volume 138, #10, pp. 1317–1361

BIOGRAPHICAL SKETCHES

Judith S. Bellin is professor of biochemistry at the Polytechnic Institute of New York, where she teaches biochemistry and environmental chemistry. She organized a group of colleagues and students to investigate the health of pesticide formulators, and helped them achieve better working conditions. As a member of the field investigation team of the Environmental Sciences Laboratory of the Mount Sinai Hospital, Dr. Bellin participated in field investigations on the health of battery workers, and of Michigan farmers and their families, who were exposed to polybrominated biphenyl flame retardant chemicals. Dr. Bellin is currently assigned to work with the Environmental Protection Agency in Washington, where she is helping to develop regulations to control hazardous wastes.

Susan Bram, Ph.D. is the Coordinator of Child & Adolescent Psychology, Assistant Professor of Psychology, and Assistant Attending Psychologist at New York Hospital-Westchester Division, Cornell University Medical College. She has previously worked at city and state hospitals in the New York area, treating adults and children on both an inpatient and outpatient basis. Her research interests include cross-cultural aspects of psychology, sex roles, and the psychology of reproduction. She received her Ph.D. from the University of Michigan in 1974.

Dr. June Jackson Christmas is Medical Professor and Director of the Program in Behavioral Science at the School of Biomedical Education, City College, City University of New York. A certified psychoanalyst, Dr. Christmas is a graduate of Vassar College and received her medical degree from Boston University. She was awarded honorary degrees by Boston University and Trinity College. From 1972 to 1980 she served as Commissioner of Mental Health, Mental Retardation, and Alcoholism Services for the City of New York. From 1979 to 1980, she served as President of the American Public Health Association; she has also served as Vice-President of the

American Psychiatric Association. June Christmas headed the Carter/Mondale Transition Planning Group developing policy options and initiative for the Department of Health, Education, and Welfare, and was the Executive Coordinator of the Task Panel on Community Support Systems of the President's Commission on Mental Health. Earlier, Dr. Christmas established and directed the Harlem Rehabilitation Center. In 1976, she received the American Public Health Association Award for Excellence.

Myra Fooden is a psychologist in private practice in New York. She has held faculty positions in departments of Psychology and Education in City University, Long Island University, and Empire State College-SUNY. Dr. Fooden is active in several professional groups concerned with the status of women and has lectured and conducted workshops in professional and public settings. She received her PhD degree from Yeshiva University in 1974.

Victoria Hashmall Freedman was born and bred in New York City. She attended the New York City public schools and graduated cum laude from Brooklyn College. After a brief stint as a research assistant in the Department of Medicine at New York University School of Medicine, she entered the graduate program at the Albert Einstein College of Medicine (Bronx, N.Y.), where she pursued her interests in genetics and cell biology. Her doctoral research focussed on the parameters of cellular malignancy in order to identify the characteristics of cell growth in vitro specifically associated with the ability of cells to generate tumors in experimental animals. Currently, she is at the Rockefeller University in New York, where her research concerns the delicate immunological balances involved in the growth of tumors, and the manipulation of particular cell populations to combat tumors. When not in the lab, Dr. Freedman's time and interests are totally absorbed by her husband and two lively young daughters.

Marji Gold received her M.D. from New York University Medical School in 1973. She is currently on the attending staff and a member of the faculty of the Department of Family Medicine, Montefiore Hospital. Her responsibilities include teaching residents, patient practice in the Bronx and coordinator of curriculum in obstetrics and gynecology for family medicine. She is also attending physician in the High Risk Obstetrical clinic at North Central Bronx Hospital. She is a member of the International Committee Against Racism.

Susan G. Gordon received an M.D. degree from Howard University in 1950. She is presently an Associate Professor of Clinical Pediatrics at Columbia University having spent 25 years teaching and practicing pediatrics. She served as an attending Pediatrician at Columbia Presbyterian's Babies Hospital, St. Lukes-Roosevelt Hospital, and Harlem, Metropolitan and Flower-Fifth Avenue Hospital. From 1952–56 she served as co-director with her husband Edmund W. Gordon of the Harriet Tubman Child Health and Guidance Clinic in New York City. She is the mother of 4 children.

Ben Harris, a clinical psychologist, received his B.A. from Hampshire College, Ph.D. from Vanderbilt University, completing his internship at Baylor College of Medicine. He has been an Assistant Professor of Psychology at Vassar College since 1978, and is spending 1982–83 as Visiting Scholar in the University of Pennsylvania's Department of History and Sociology of Science. He is a member of a variety of professional organizations, including the Group for a Radical Human Science. In 1982 he was appointed to the National Citizen's Commission to Investigate the Marion (Ill.) Behavior Control Unit.

Betty Rosoff, Ph.D. from the City University of New York in 1967; endocrinologist and physiologist; Professor of Biology, Stern College of Yeshiva University. She is a member of the International Committee against Racism and is on the editorial board of the journal The Prostate. She applies her expertise in endocrinology to the investigation of gender and biological determinism.

Reva Rubenstein is a board-certified toxicologist. She taught occupational health and toxicology to labor groups and allied health professionals. She has worked as a consultant for trade unions. She is presently working for a trade association concerned with chemical waste.

Jagna Wojcicka Sharff, Ph.D. Department of Anthropology, Columbia University, 1979; ethnographic research among diverse urban ethnic groups, the underground economy and adaptive strategies to poverty; member Community Relations Committee of the American Friends Service Committee; teaching Medical Anthropology at City College in New York.

Ethel Tobach, Ph.D. from New York University, 1957, comparative psychologist at the American Museum of Natural History and Adjunct Professor in Biology and Psychology at The City University of New York; research on the evolution and development of social-emotional behavior; has written extensively on the role of science in societal processes leading to racism and sexism.

Georgine M. Vroman, Ph.D. Graduate Faculty, New School for Social Research, 1979; medical anthropologist. Grew up in Indonesia and studied medicine at Rijks Universiteit, Utrecht, Netherlands; has lived in the U.S. since 1947. At first participated in biomedical research, then did extensive volunteer community work while raising a family, eventually returned to graduate studies. Special concerns are rehabilitation of brain-injured patients, aging, urban issues, the interface between biology and culture. She taught Environmental Health at Ramapo College, New Jersey, and, at present, works at the Cognitive Rehabilitation Unit of Bellevue Hospital Center, New York City.